The Fossil Hunter

The Fossil Hunter

Dinosaurs, Evolution, and the Woman Whose Discoveries Changed the World

SHELLEY EMLING

 ST. MARTIN'S GRIFFIN ❦ NEW YORK

ISBN: 978–0–230–10342–9

Library of Congress Cataloging-in-Publication Data

Emling, Shelley.
    The fossil hunter : dinosaurs, evolution, and the woman who changed
science / Shelley Emling.
        p. cm.
    Includes bibliographical references and index.
    ISBN 0–230–61156–7
    (paperback ISBN 978–0–230–10342–9)
        1. Anning, Mary, 1799–1847. 2. Women paleontologists—England—
Biography. 3. Paleontologists—England—Biography. 4. Discoveries in
science—History—19th century. I. Title.

QE707.A56E46 2009
560.92—dc22   2009017900
[B]

A catalogue record of the book is available from the British Library.

Design by Newgen Imaging Systems (P) Ltd., Chennai, India.

First ST. MARTIN'S GRIFFIN paperback edition: January 2011

P1

*To Chris, Ben, and Olivia, who make
everything—everyday—better*

Contents

Acknowledgments

Since her death, Mary Anning has garnered a small but hard-core group of fans who have spent countless hours of time researching her life. Without them, it would be impossible for someone like me to tell her story in my own way. William Lang (1878–1966) was Keeper of the Department of Geology at the British Museum, and, after retiring to Charmouth, he gathered together a wealth of information on Mary's work and background. Hugh Torrens is a geologist and historian who also has done an enormous amount of research on Mary, and was kind enough to meet with me. Other writers and historians who have done incredible work on Mary and other scientists of her era include Thomas W. Goodhue, John Fowles, Deborah Cadbury, Patricia Pierce, Christopher McGowan, and Paul J. McCartney.

Personally I'd like to thank my agent, the tireless Agnes Birnbaum, who never, ever gave up. Thank you to my incredible editor at Palgrave Macmillan, Alessandra Bastagli, who believed in—and delighted in—Mary's story. Also thanks to Colleen Lawrie at Palgrave for kindly helping me through all the nitty-gritty of delivering a book.

Thanks to the many friends and family members—especially Carolyn, Ginger, Paula, and Special K—who have always taken an interest in my progress on the book. Thanks also to Tom and Annette Buk-Swienty for inspiring me and to Peter and Rachel Hayward for kindly allowing me to stay in their lovely home—a great base for visits to Lyme Regis. Thanks to Sasee Cats—you know who you are—for the daily laughs. Let's hope they never end. Anything I manage to accomplish is dedicated to the memory of my wonderful mother—always. Finally, my biggest debt of gratitude goes to Scott Norvell, who must be the most supportive husband on the planet.

I only hope this book will make you appreciate Mary Anning's extraordinary talents the way I do. I have taken a few liberties in an effort to fill in some gaps in Mary's life but tried to make it clear when this was the case. It is incredible that so many people have never heard of this great woman. Although many children's books have been written about her accomplishments as a young girl, less attention has been paid to her string of discoveries as an adult. Not only was this uneducated woman able to hold her own with some of the best minds of Europe, but she also displayed an amazing ability to both ferret out and restore fossils. Perhaps this book will go at least a small step toward introducing Mary to more people. That's my only wish. Enjoy!

Prologue

She sells sea-shells on the sea-shore
The shells she sells are sea-shells, I'm sure
For if she sells sea-shells on the sea-shore
Then I'm sure she sells sea-shore shells.
—Tongue-twister written by Terry
Sullivan in 1908 and inspired by Mary Anning

*P*re-Victorian England exemplified a powerful period in
the history of science, a time when one never-before-seen
monster after another was being cajoled from its Jurassic tomb,
drawn out into the light of day where it could blow holes through
the Biblical account of the earth's history. These creatures—with
their bat-like wings, snake-like necks, and big, bulging eyes—vividly brought to life a prehistoric era that was more bizarre and
harrowing than anyone had ever imagined. Indeed, they forever
changed the way people thought about the world around them.

Today it is hard to relate to the mindset of people living in the
early nineteenth century, decades before dinosaurs burst onto
the scene. Without question, most people accepted the idea of an
earth created in six divinely ordered twenty-four-hour time slots,
in 4004 BC—a year derived at by meticulously tracing the biblical
genealogies. After creating the sun, moon, stars, and oceans, God
made creatures of the air, sea, and land on day five, followed by
the appearance of the most magnificent and most complex creature—man—on day six. The land animals harmoniously shared
the garden with Adam and Eve and all were vegetarian until
the first couple sinned and thus began meat-eating and mortal-
ity. Later, during Noah's time, the global flood decimated all life
except for that which had been corralled into his ark. Fossilized

seashells found on mountaintops were proof positive of a deluge so great it swept up and over everything in its path.

In the early 1800s, most people had absolute faith in the fact that species never changed or evolved, or became extinct. Everything that existed had always existed. The world was pretty simple, really. There wasn't any radioactivity or relativity, extinction or evolution, to muddle things up. It was during this time that a young working-class English girl named Mary Anning began raising eyebrows with her daily forays along the dangerously crumbling strata of England's southern coastline, decked out in voluminous tattered skirts as she ran the gauntlet of high tides and landslides. Always she was on a hunt for curiosities she could sell to seafaring tourists in order to put food on her table.

By birthright, Mary should never have grown up to be a famous fossil hunter and geologist. In addition to being dirt poor, Mary Anning also was marginalized by odds clearly not stacked in her favor: her sex, regional dialect, lack of formal education, and adherence to the Dissenter faith, a religious strain that didn't conform to the teachings of the established Church of England.

But she enjoyed one powerhouse advantage: the very good fortune of having been born in exactly the right place at the right time, in 1799 in an unassuming English town called Lyme Regis alongside some of the most geologically unstable coastline in the world. Unbeknownst to anyone at the time, its wobbly cliffs held the remains of a baffling array of ancient reptiles, reptiles that used to roam the land and inhabit the seas hundreds of millions of years in the past.

I first discovered Mary Anning a few years ago during a vacation to England's so-called Jurassic Coast with my husband and three children. My ten-year-old son Ben is quite the budding geologist who likes nothing more than to wait for low tide, then scramble out across the shore below vertical, yet slumping, cliffs, hoping to chance upon an ammonite in a grayish stone streaked with quartz. During our brief trip, we paid a visit to the small Lyme Regis Philpot Museum where, tucked away on the second floor, was a permanent exhibit designed to preserve the legacy of a woman I'd never heard of—Mary Anning.

It wasn't so much Mary's fossil finds that intrigued me, but rather it was how this low-ranking woman had managed to make an indelible mark in such a male-dominated field. I was bowled over by how, in 1823, she successfully challenged Georges Cuvier—perhaps the most influential figure in science at the time—following her discovery of a plesiosaur. But most remarkable of all was that almost no one I knew had ever heard of Mary Anning—even among my British friends. And this was a woman London's Natural History Museum refers to on its website as the "greatest fossil hunter ever known."

Not only were Mary Anning's finds extraordinary, but so too was the fact that she was out of step with her times, a rebel who carved out her own niche in a highly stratified and sexist society. All around her were learned men who were trying to reconcile the gap between religious beliefs and scientific evidence—and they were using her fossil finds to do it. Indeed it was Mary's spectacular marine reptiles that pushed them into finally contemplating a different explanation for the world's origin.

Mary's many finds also laid the groundwork for Charles Darwin's theory of evolution, elucidated in his 1859 *On the Origin of Species*. Darwin drew on Mary's fossilized creatures as irrefutable evidence that life in the past was nothing like life in the present.

Mary was deeply pious and—like Darwin—she most likely was unsettled by what she was unearthing. She couldn't have possibly imagined where her finds would take her, and eventually Darwin, and the rest of the world. Always, though, she maintained a belief in God's omnipotence. Most amazingly, this girl, with dirt under every fingernail, with barely a shilling to buy vegetables, grew up with so much confidence that in 1844 she was able to brazenly tell the King of Saxony's entourage: "I am well known throughout the whole of Europe."

Here is her story.

I

Snakestones, Thunderbolts, and Verteberries

On a sunny September afternoon in 1793, an adventure-seeking cabinetmaker named Richard Anning and his brooding bride, Molly, packed up a few trunks of belongings—all they owned in the world—and journeyed about a dozen or so miles, from the tiny village of Colyton, to England's dangerously rugged coastline. They had decided to stake their future on an obscure port town they'd probably heard of in passing only a few months before: Lyme Regis.

Most likely equipped with little more than an old family Bible, farm tools, a few cotton skirts and shirts, and a single waistcoat, the young couple might have paid about 10 shillings to hitch a ride on a large, heavy wagon used to ferry equipment to farms dotting the bucolic borough of Dorset. Drawn by eight horses, the vehicle would have snaked along at a tedious two miles an hour toward what Richard felt sure in his heart would be a prosperous future.

By most descriptions, Richard Anning was a charismatic, somewhat childlike man with a passion for change and new challenges; his wife was much more serious and cynical. Richard's faith was in the Lord and in his own exhaustive energy. Molly's trust was in Richard.

After nearly two days of weary plodding, across rocky streams and bright meadows, bounced about among piles of saddles and harnesses, the bedraggled newlyweds would have heaved a huge sigh of relief when they finally caught sight of a few tightly packed houses on a hilly road, tumbling down toward a delicious little harbor: the first sight of their destiny.

Mingling with the more common thatched cottages and stone barns was a sweet array of white Georgian villas, somewhat faded by decades of salt spray, but still providing a rare bright spot in what was otherwise a sobering picture.

For the most part, Lyme Regis looked like an unnatural town, squashed as it was into the narrow valley of the river Lym, book-ended by miles and miles of England's most unstable coastline. With space ideally suited for a population of 800, the bowl of a village had become home to at least 1,250 mostly hard-up residents by the time Richard and Molly Anning rolled in. As a result, sanitary conditions were deplorable, with the town's compactness contributing to the rapid spread of dirt and disease. Hogs and rats ran amok in the streets. And, at house after house, mounds of refuse rose up from the ground like sunflower plants, towering so high that it was hard to see out the windows. Just that summer, the Lyme courts had prosecuted a handful of citizens for allowing bullocks to defecate into clean drinking water. Indeed, the water was so filthy that many people bypassed it altogether, opting instead to quench their thirst with a watered-down ale known as "small beer." Casting an invisible pall over the town were rumors that France, the dominant military power in Europe, was about to launch an invasion. Lyme Regis was right in Napoleon's line of fire.

This was the new world—the great destiny—that greeted Richard Anning and his bride as they stretched achy legs, perusing a landscape that exuded commotion but hardly the promise that should have gone with it. Most wives would have been horrified. In the most positive of lights, it was a place that would take some getting used to, even for the hardiest of characters. But where Molly might have seen barriers, cheerful Richard would have seen only opportunities. If anything, his spirits would have been buoyed just by the change of scenery. And—whether it was intelligence, intuition, or just dumb luck—Richard's instincts would turn out to be flawless.

Richard's choice of Lyme Regis had been a strategic one, most likely triggered by news from friends or travelers of a new turnpike built between the nearby towns of Dorchester and Exeter—a road completed only in 1758—that was designed to pump economic life into the entire region.[1] Until 1759, a wheeled vehicle had never passed into Lyme Regis; all goods landed at the port and were transported by packhorse teams. The borough of Dorset also was marginalized from the rest of the country because of its very strong local dialect. The turnpike helped change all that by linking Lyme Regis to the outside world. Probably unbeknownst to Richard when he arrived was the fact that—with the help of the new turnpike—Lyme Regis was about to transform itself into England's guilty little pleasure.

For centuries, the town had been a major commercial port. Even in the late 1700s, it was one of the most important ports in England. In particular, Lyme Regis had taken enormous pride in its manmade breakwater—one of the oldest artificial harbors in England, dating from the thirteenth century. Known as the Cobb, which means "rounded island" in the local dialect, this 600-foot stone jetty, built from oaken crates topped with rounded boulders, curved out protectively like a long arm into the English Channel, shielding the location against fierce westerly storms and buffeting seawaters since medieval times. But ships were becoming too large for the town's shallow harbor, with its limited space, and so fewer were arriving each month. Once-busy shipyards near the Cobb started going out of business, and over time, most trade transferred from Lyme Regis to Liverpool, a port about 250 miles away. The same ships that couldn't carry goods out also stopped bringing food in, resulting in widespread food shortages. Adding impetus to this economic downturn was a disappointing cloth trade, slammed by highly organized competition from a more industrialized North. With its plethora of weavers' homes and mills, Lyme Regis had enjoyed a booming cloth business with France as early as 1284. But the Industrial Revolution, which started in England in the mid-1700s, meant that steam-powered looms in cities farther north started churning out cloth and lace faster and cheaper than they could be crafted by hand. Coupled with war in Europe, the Industrial Revolution hammered the town's longtime economic staples.

By the early 1700s, prophecies of economic doom were becoming commonplace, and many families were barely able to put food on their tables. But desirable new prospects were arising. Lyme's salvation turned out to be its clean sea air and salubrious seawater—perks the locals had long taken for granted but that others in England, weary of the grinding noise and filth brought on by the Industrial Revolution, were now seeking with a voracious appetite.

And so the Annings were but a few drops in a stream of English travelers who were starting to discover Lyme Regis during those days, their journeys more often than not inspired by one Richard Russell, a prominent doctor from Brighton. Russell had created a sensation in the 1750s, writing a popular treatise on the therapeutic effects of the ocean. He called it "The Dissertation on the Use of Seawater in the Affections of the Glands."[2] The therapy was simple enough: complete immersion in the sea followed by consumption of a pint of seawater, a routine he described as "a common defense against corruption and putrefaction of bodies." Soon the seawater was considered a cure for everything from gout to gonorrhea. The storm of publicity surrounding Russell's work tempted affluent tourists to travel to England's coastal villages. Even King George III, who oversaw the loss of the American colonies, was known to take long dips in the sea during retreats to Weymouth, near Lyme Regis, with the robust strains of "God Save Great George Our King" blaring out from musicians on the beach each time his portly frame descended into the water.

Almost overnight, and just at the tail end of the eighteenth century, Lyme Regis had turned into a spa town, with its local government officials and business owners cleaning up the streets and promoting the shoreline as an economic alternative to renowned but more expensive rivals such as Weymouth, Brighton, and especially Bath. With its warm mineral springs, Bath had pioneered this new fad, although its perfect blend of entertainment and architecture had made it almost too successful, and so crowded that pleasure seekers began looking for quieter alternatives. In general, English travelers deterred from visiting the Continent due to tensions with France were eagerly seeking holiday options that were closer to home.

Coastal towns like Lyme Regis also were benefiting from the invention of a new essential for enjoying the sea: the bathing machine. Previously, anyone wishing to savor the water was obliged to undress on the beach. The new contraption, although cumbersome to use, was designed to provide cover for genteel souls for whom modesty was next to godliness. Donkeys drew to the water's edge what was essentially a wooden shed on wheels, providing a sort of hidey-hole for the bashful. By the early 1800s, four bathing machines were operating on the western part of Lyme beach, with another seven regularly plying the tranquil waters behind Lyme Regis's manmade harbor. Men and women took turns using the machines. When ladies bathed, a bell was sounded, meaning men should make themselves scarce. Even rowboats were requested to stay at least 100 yards away.

It wasn't long before indoor baths became popular as well, with one of the first being built in Lyme Regis in 1804. These baths boasted not only private cubicles and tepid water, with attendants, but newspapers, refreshments, and even card-game tables. They were seen as a real luxury for those who preferred not to bathe in the sea, out in the open. For nighttime entertainment, the town's chief public space, the Assembly Rooms built in the late 1700s, became the place to be, the scene of fancy balls, card games, and billiards. According to resident novelist and historian John Fowles, who set his classic book *The French Lieutenant's*

Woman in Lyme Regis, snobbery and backbiting reigned supreme amid the dancing, gambling, and gossiping, with young flirts and imperious old ladies fraternizing with gruff sea captains and pretentious young businessmen.[3]

The town was fortunate in that it had a major benefactor at the time—the philanthropist Thomas Hollis, who helped turn Lyme Regis into a healthy holiday resort by buying land along the shore and creating the first public promenade in 1771. A social reformer, Hollis was well known throughout Europe and America. His interest in Lyme Regis sprang from his retirement to nearby Corscombe. He purchased many of the dilapidated properties in town during the mid-1700s and rebuilt them in the elegant late Georgian style. Although he never visited America, Hollis later was known for his large donations to Harvard University.

By the turn of the nineteenth century, even England's artists were starting to take notice of Lyme Regis, not because of the sea's benefits but because of the region's wild and romantic coastal backdrops. Eventually the town would become a favorite haunt of James Whistler and J. M. W. Turner, both of whom completed major paintings while on holiday there. The not-yet-famous Jane Austen was another regular visitor to the area. Born in 1775, Austen moved with her family to Bath after her father retired as the parish rector in nearby Steventon. After a coach service was inaugurated between Bath and Lyme Regis, a journey of about 70 miles, Jane's family visited the town briefly in both 1803 and 1804, reveling in what was fast becoming a carnival atmosphere. In her novel *Persuasion*, Austen famously described Lyme Regis:

> The walk to the Cobb, skirting around the pleasant little bay, which, in the season, is animated with bathing machines and company; the Cobb itself, its old wonders and new improvements, with the very beautiful line of cliffs, stretching out to the east of the town, are what the stranger's eye will seek, and a very strange stranger it must be who does not see charms in the immediate environs of Lyme, to make him wish to know it better.[4]

Austen even had an altercation with Richard Anning during one of her visits. In a letter to her sister Cassandra dated September 14, 1804, Jane says she asked him to repair a "broken lid" on a box. Apparently he said the job would cost five shillings. She thought the price completely outlandish, arguing that it was "beyond the value of all the furniture in the room together."[5]

By all accounts, Richard Anning lost Jane's business but it wasn't likely that he cared too much. By this time, he was focused on other lucrative ventures. Rather than apply himself to his carpentry work, as had been the original plan, he was busy signing on to another kind of life: the finding of odd-looking stones on the beach that he could sell to tourists seeking souvenirs of their seaside vacations.

Since the sixteenth century, people had referred to most anything ferreted out of the ground—including minerals, metals, and rocks—as "fossils," derived from the Latin word for "having been dug up." Finally, by around 1800, a few academics, mostly in Paris, began to study plant and animal fossils scientifically. The strange fossils found along Lyme Regis's shores had baffled the locals for as long as anyone could remember. They came in all forms and sizes—including what later were determined to be bivalves, ammonites, belemnites, and brachiopods—and sometimes even the fragments of giant critters never heard of before. For example, what locals called "verteberries" or "crocodile teeth" actually were individual vertebrae from an unknown prehistoric creature. Lyme Bay's beaches had earned a reputation for being so full of extraordinary fossils that smugglers running ashore on pitch-black nights were able to determine their location simply by running a handful of the peculiar pebbles through their fingers.[6]

Some people thought fossils were so lovely and delicate that they surely must be God's decorations, allowed to bubble up from the inside of the Earth, a bit like flowers, plants, and trees were allowed to ornament the outside. Others thought they must be the remains of the victims of the worldwide flood recorded in Genesis. The belemnite fossil—known by locals as the "thunderbolt"—looked like an elongated bullet with an exaggerated point and was thought to be the result of a lightning strike. Indeed, a myriad of myths and theories grew up around these and other fossils. According to one, powdered belemnites were able to clear up infections in a horse's eye; water in which belemnites had been soaked was thought to be able to cure horses with worms. Called snakestones—or "cornemonius" in the local dialect—ammonites were believed to be particularly magical, able to ward off all sorts of maladies, from snakebites to impotence.

Set in the center of a fossil-rich coastline extending 95 miles from Exmouth to Bournemouth, Lyme Regis's stretches of beach were and still are some of the most unstable expanses in the country.[7] The cliffs' limestone ledges alternate with bands of shale, all crazily tilting toward the sea at some point, like a geological deck of cards. The mix of limestone and shale is formally known as Blue Lias—blue for the blue-gray color of the limestone and lias from the Gaelic word for "flat stone." This Blue Lias was, and is, constantly under siege from the rough waves. Every time a cliff face withers away, eroded by water, a rich seam of fossils is laid out, as nicely as a Sunday supper. But a fierce storm can just as easily wipe out and wash away a unique fossil before anyone can get to it.

Many of Lyme Regis's most important treasures have always been tucked away tightly inside Black Ven, the ominously named 150-foot-high cliff of sliding sand topped with scrubby grass, located east of town. Countless other strange stones are hidden inside the adjacent Church Cliffs, also to the east, where the limestone and shale could be reached only after the tide receded. Much later, geologists would learn that Blue Lias belonged to the lowest division of the Jurassic—part of the Mesozoic Era and also known as the "Age of Reptiles." During this era, some 200 million years ago, this region had supported diverse life-forms. At the bottom of the food chain were plankton as well as clams, oysters, and barnacles, sources of food for other sources of food that eventually reached to the top of the food chain—to carnivores and various sea scavengers.

At the time, most of Britain was covered by generally shallow waters with mountains rising up from some parts. The site on which London would be built was located on a peninsula, overtaken by rising water levels. What is now America was drifting away from Europe with the Atlantic starting to take shape as a recognizable ocean. As time passed, all sorts of life-forms were safely deposited in ancient seabeds, turning Lyme Regis's topography into something of a primordial soup, a place uniquely able to store evidence of 200 million years of evolution. Scientists eventually discovered that the cliffs east and west of Lyme Regis portrayed an almost continuous sequence of rock formations spanning the entire Mesozoic Era, perhaps better than any other locale on the planet. Until the early 1800s, though, the area's residents had no knowledge of this. But then, ever so slowly, the nearby cliffs began spilling their secrets, opening up like the cover of a prehistoric tome sealed away by the eons.

At high tide and at the height of any potent storm, Lyme Regis's coastline became a sort of Ground Zero for unimaginable violence. A strong northeasterly wind carried waves from halfway across the North Atlantic, water that crashed into fragile walls of mud, clay, shale, and limestone, exposing at the rate of two to three feet a year what were millions of years' worth of geological truths. Every so often these storms brought to light proof that ancient monsters had once roamed the land: a dimming outline of a sharp tooth on a flat stone; an impression of a knee joint peeking out from under a clump of weeds. Petrified in the rocks were all sorts of bizarre-looking fragments, hints of a lost world in which the dinosaur was king.

Yet the Lyme Regis fossils remained nothing more than curiosities to the people who found them and to the tourists who bought them. Most of the fossil hunters were interested only in their earning potential. And few would have understood these earning possibilities better than the industrious Richard Anning.

Richard and Molly settled into a sturdy but primitive timber-framed house with wood paneling and three windows facing the square and many smaller windows facing the sea. They were able to rent the place at an especially low price as the modest edifice sat smack next to the town's jail, the Cockmoile. A nuisance at best, the local hoosegow drew unsavory characters, the kind that made most respectable folk avoid walking by after dusk. Cockmoile Square was oddly shaped and densely packed, anchored in one corner by the town hall, which shared a building with the butcher shop. The square also was home to the stocks, used until 1837. Next to the Annings were the Bennetts, a family they apparently got on with. John Bennett was the shoemaker who later opened his own private baths for tourists while his wife, Maria, looked after their daughter Ann. Nearby was the town's most important hotel, the Three Cups Inn.

Most vexatious for the Annings were the two narrow roads that ran right outside their door, a blind intersection considered among the most dangerous in England, infamous for its high number of accidents. Rarely did a week go by when a wagon didn't collide with someone pushing a wheelbarrow. But Richard Anning rarely complained; unlike his rather dour spouse, he wasn't a type prone to bleak thoughts.

A bearded giant of a man, Richard had grown up about a dozen miles from Lyme Regis, in a medieval village called Colyton, believed to be the most rebellious town in the borough of Devon.[8] Indeed its citizens were the first to resist the high taxes imposed by King Charles I, with many eventually executed for treason. But it was a pretty town too. Nestled in Devon's rolling countryside on the banks of the river Coly, the small town of Colyton was set along a circular Saxon pattern of winding streets. These were decked out with the most remarkable range of trades: bakers, brewers, tanners, farmers, milliners, gunsmiths, shoemakers, watchmakers, stone masons, and market traders. Richard Anning came from a long line of cabinetmakers, accustomed to working with their hands. Although he became a skilled craftsman, carpentry work might have been stifling for a spirited personality like Richard, and his ambitions never seemed to settle down. Always, he yearned for more.

He just needed a dependable partner for whatever new enterprises awaited him. Perhaps at a church social, the serious-looking but not unattractive Mary "Molly" Moore caught his eye. Molly had grown up as poor as Richard under the loving but watchful eyes of her parents in the nearby sleepy hamlet of Blandford. She seemed to be a hard worker and smart, a bit timid and insecure, but quickly accepting of the charming Richard Anning's attentions and eventually his proposal. In truth, she probably never had been overwhelmed with suitors.

Not even a month after their wedding, held at Blandford parish church on August 8, 1793, Richard announced that they were moving several miles away to Lyme Regis. Despite the fact that they might as well have been putting down roots in a foreign country, far from everyone and everything she had ever known, Molly probably wasn't in a position to object. Like most new brides living in a male-dominated era, she would have had to follow the whims of her husband.

Once settled into their new home, Molly quickly recognized that Richard was investing a lot more time in beachcombing than he was in carpentry work. Within only a few weeks he'd set up a round table outside their cottage, laden with fossils, hoping to catch the eye of any passing tourists. Each small sale goaded him to seek the next. Most likely Molly thought it irresponsible for the head of a household to focus on a field in which he had no track record.

But as the years passed, Molly faced more pressing concerns than coping with a headstrong husband. Although Lyme Regis was becoming a sweetly soporific place to visit, it was a frightfully harrowing locale for any mother trying to keep her children alive. The Annings' union had first produced a daughter, Mary, in 1794, followed by a son, Joseph, born in 1796. Several other children soon followed. Across England at the time, women could expect to lose more than a quarter of their offspring before their first birthday and another quarter before their fifth. The causes for this high death toll included pneumonia, smallpox, and measles, all as common as colds but much more fatal.

Lyme Regis offered an additional challenge. The coastline on which the town sat was, statistically speaking, one of the most perilous locations in England, and many families would never have dreamed of raising children there. The same vicious storms responsible for eroding the cliffs to reveal fossils were also known to drive huge waves up the river Lym and straight into Richard and Molly's home, one time nearly sweeping away the entire first floor. Once, the family was forced to crawl out of an upstairs bedroom window to escape drowning.

In addition, the Napoleonic Wars were causing severe food shortages. The ambitious French emperor had stationed 10,000 soldiers across the Channel at Boulogne, which forced Britain to ready its own army. With no European corn on the market, the price of English wheat rose sharply, from 43 shillings a quarter (eight bushels) in 1792 just before the war to 126 shillings a decade later.[9] Before long, many of Dorset's poor were spending half their incomes on bread. At the time, many eked out a living on only about six or seven shillings a week. Stagnant wages weren't much help, forcing many people to seek supplements from their local parishes. Even the most diligent of laborers realistically feared starvation. Meanwhile, the well-off gentry, sealed off inside their sprawling country estates, seemed impervious to the ravages of war as they reaped the rewards of the higher prices. But riots were erupting up and down the coast as flaming barns became the new calling cards of the increasingly frustrated poor.

In December 1798, the days were so cold that everyone in town cared only about one thing: keeping warm. Due to constant shortages of firewood and coal, the poor already had endured many months of damp, icy weather. But that December brought with it some of the most frigid days on record. Just a few days before Christmas, Mary, the Annings' oldest daughter at the time, was playing with her brother Joseph in a room where Richard had stored some wood shavings. Molly left the children alone not even five minutes—just long enough for the youngsters to toss a few more shavings into the fireplace, probably in an innocent effort to feed the fire. In seconds, the little girl's clothes went up in flames;

the four-year-old was horribly burned as her two-year-old brother looked on. By the time Molly returned, little Mary Anning—the first child named Mary in the family—had stopped breathing and could not be revived.

The tragedy was so horrific it wasn't even mentioned in the local paper. But *The Bath Chronicle* reported the incident: "A child, four years of age of Mr. R. Anning, a cabinetmaker of Lyme, was left by the mother for about five minutes...in a room where there were some shavings...The girl's clothes caught fire and she was so dreadfully burnt as to cause her death."[10]

Most likely, any woman in Molly's position would have turned angry, depressed, remorseful, and guilty, certain she was a failure as a mother; and by all measures this was one of the roughest, bleakest periods in Molly's life. Adding to her pain over the years was the loss of two more children, Martha and Henry, in the few years after Mary's death. Eventually Molly would bury eight of her ten children. It's hard to know what kind of support Molly might have received from friends. What is likely is that the family would have been looked down on, at least by some of their neighbors, as a result of their religion.

Certainly southern England had long had a tradition of religious nonconformity. After a succession of rulers ordered funds to be taken from the Customs at Lyme to maintain the Cobb jetty, Catholic Queen Mary withdrew the grant because the "inhabitants were then reputed as heretics for their religion."[11] Yet across the country, even in southern England, few labels held a more negative connotation than Dissenter, with the word being used to describe anyone who failed to follow the state-supported Church of England, known as the Episcopal Church in the United States, regardless of whether that person was Presbyterian, Baptist, Quaker, or the adherent of some other Christian denomination. Until 1689, it was illegal to be a Dissenter, and even after the law was changed, in 1828, a Dissenter wasn't allowed to join the army or to hold any official position. A Dissenter wasn't allowed to attend university and was prohibited from entering several professions.

Soon after arriving in Lyme, Richard joined the Independent Chapel on Coombe Street. Here a group of Dissenting worshippers called themselves Independents—later they'd be known as Congregationalists. In the late 1700s, Independents made up one of the largest bodies of worshippers not associated with the Church of England. But Richard stood out even among nonconformists. He raised the eyebrows of other Dissenters by fossil hunting on Sundays, Good Friday, and other holy days. Every time he took his small children with him to the beach to help him hunt for curiosities, where a rising tide might have trapped them, he was the focus of much gossip. Molly might have felt the sharp sting of ignominy. But, unlike his wife, Richard seemed to care not a whit about what anyone else thought.

On May 21, 1799, Molly gave birth to a tiny baby she named Mary, in memory of her other daughter, Mary, who had died in the fire only five months earlier. Following what was likely an easy labor, the smallish baby stubbornly clung to life although she remained in precarious health.

Little Mary was about 15 months old, sickly and weak, eating little, crying buckets, and coughing constantly, when a peculiar event transformed her life. In those days, any sort of diversion from the drudgery of routine chores was welcome. On the third Sunday in August of 1800, the diversion was a touring company of riders who had camped in Rack Field on the outskirts of town and promised an entertaining show on the weekend.[12] Elizabeth Haskings, a sweet-natured local nurse who sometimes helped Molly with the children, offered to take little Mary to the outdoor equestrian show so that she could get some fresh air. Molly was grateful for the respite. And so Elizabeth and Mary, and most everyone else in town, trotted off to watch the vaulting and riding stunts and, more important, to buy a shot at a copper tea kettle or leg of mutton in the raffle. In sharp contrast to the tattily dressed locals, the men on horses probably whipped by in elegant double-breasted cutaway jackets with long tails and riding breeches. All in all, it would have been a splendid performance.

But by late afternoon, as the show was building to its climax, large cumulonimbus clouds seemed to form out of nowhere from behind the western hills. Soon a vicious storm, a rare electrical one, began to roll in over the small seafront valley. Still, the crowds wouldn't budge. They were eager to hear who had won the lottery. One of the onlookers, the local schoolmaster and historian George Roberts, later recalled that it had been just after 4 P.M. when a thunderclap broke the stillness and echoed around the cliffs of Lyme Bay. "All appeared deafened by the crash," he said. With the boom came a blinding shot of bright light followed by dead silence. But in a moment the scream of a grown man shattered the quiet. He was pointing to a group of women lying motionless under a lofty elm tree, their white frilly dresses looking like silhouettes against the thick green grass.

Among the women was Elizabeth, much of her hair singed into a blackened mat and her entire right side charred. The lightning had entered the group through the tree's roots and then arched back up the legs of Elizabeth and her two friends. The tree had shattered in place, littering the ground with confetti-size pieces of bark, and the victims' clothing had been blackened around its lacy edges. Like firecrackers, one woman's shoes had exploded. There was nothing anyone could have done.

But Elizabeth still gripped baby Mary, who appeared to be breathing faintly. Maybe, thought those in the crowd, she could be revived. Someone from the horde scooped Mary from Elizabeth's rigid arms and rushed her to the Annings' home at the center of town, a mass of townspeople hot on his trail. There, Mary's parents dunked their precious girl in a tub of warm water and sent for a doctor. Most likely the stench of burned hair and flesh and the memories it rekindled would have been nearly unbearable for Molly. Somehow, in some way, Mary's color returned and her breathing steadied. The local physician, Thomas Carpenter, declared it a miracle to which the assembled crowd let out a cheer so loud it was nearly heard all the way to Lyme Beach.[13]

For Mary, and for her mother, Molly, the lightning strike marked a turning point. Many years later, Lyme Regis residents would say that a "dull child" had been transformed into a lively, healthy, and exceedingly curious young girl by the accident. The toddler with the poor appetite now ate anything put in front of her. No one who knew the family could get over the difference.

Perhaps it was the miracle, perhaps the brightness of the child, or maybe the kindness of those in town after the incident. Whatever the reason, Mary became the source of unutterable joy for her parents and, for the first time in a long time, Molly had a reason to feel good about something.

Within a few years, when she was only five or six, Mary Anning became her father's constant companion. Richard saw great promise in his inquisitive little sidekick and often lured her from household duties to rummage the beach with him for curiosities. Never mind that this wasn't what little girls were supposed to be doing. In this era, girls of all classes were supposed to be at their mothers' sides, raised to be well-mannered adult women who heaved their bosoms at any able-bodied man in britches. Molly would have been savvy enough to realize that, without a proper husband, Mary was destined for farm work, domestic service, or a menial job in one of the many factories that were opening up as a result of the Industrial Revolution. Women had very little standing in society—particularly lower-class women from a family of Dissenters—unless they married up.

But, like her father, young Mary harbored a stubborn streak. And she craved adventure. She loved being with Richard, whether it was clambering over cliffs and laughing out loud, or cleaning stones in his dusty workshop in silence. A quick study, she also learned how to deal with customers as she watched Richard charm men and women completely out of his social league. She understood how to increase the value of a find by using a tiny brush and needle to clear away the rock that encased it. By the time Mary was six, Richard was spending hours each week demonstrating how to scrub and polish a rock's surface until its colorful bright tints shone through.

Until Richard began working with fossils, apparently no one had ever sliced ammonites in half in order to expose their brilliant crystal-lined chambers. Thanks to his carpentry background, Richard was able to show Mary how to carve intricate boxes out of wood so that people would have a place to display and store their best stones. Eventually Richard fashioned a small pick to fit the child's hands so Mary could probe away at the cliffs herself, without his help.

The Annings weren't the only ones to scour the beach for fossils. In 1805, when Mary was six, four sisters of a well-to-do London attorney named Mr. Philpot moved to Lyme Regis, purchasing a large home, called Morley Cottage, toward the top of Silver Street with a sweeping view of the town and sea. Three of them—Mary, Margaret, and Elizabeth—would live in Lyme Regis all their lives and develop into highly respected fossil collectors. In time, one sister in particular, Elizabeth, would join Mary Anning on the beach almost daily, despite Elizabeth's higher social status.[14]

That same year, torrential rain drenched southern England for weeks on end.[15] This kind of weather always sent Richard into a tailspin as he anxiously awaited his next chance to return to the beach. During a particularly bad barrage, he might have sat in his rocking chair, listening impatiently as the waves crashed against the sea walls. By this time, he had been fossil hunting long enough to know that serious storms chewed away at the coastline, and he would have been eager to take advantage of it. As the rain began to taper off, he might have grabbed his hammer and chisel, running out the door with both Mary and her older brother, Joseph, trailing behind. The children's satchels would have swung next to their small bodies as they scurried to keep up, their heavy clogs squishing in the muddy streets.

On a typical day, the threesome would have paraded east from the center of Lyme Regis over the small grassy hill by the Church of St. Michael and up toward the crumbling road that led to the nearby town of Charmouth.[16] One after the other, they would have descended to the beach. Uncomfortably pebbly at first, the ground would have quickly degenerated into a mush of brown sand that would have stuck to the soles of their shoes, weighing down their feet a little more with each step. Sometimes they would have crossed a series of narrow limestone flats, the ground pocked with curious patterns that resembled curled-up worms, some the size of a man's fingertip, others as large as a barstool. Beyond the flats, Richard and the children would have navigated more slippery rocks, then crunched over smaller shingles mixed with brownish-greenish seaweed on the way to their final destination. Almost always, they were headed for the streaked facade of Church Cliffs, less than half a mile down the shoreline from the sinister mound of dark earth that was Black Ven.

Rain always transformed Black Ven into a treacherous mass of sticky clay. At any moment, rockslides might have killed the trio without warning. Indeed, rocks fell regularly. Some, no bigger than peas, would have bounced gently off Mary's bonnet. But larger chunks of cliff—the size of dinner plates—would have fallen too, making a loud racket as they crashed down onto piles of rocks along the shore. Richard had been nearly killed more than once in the frequent avalanches. It didn't help that the cliffs at Black Ven were among the most unstable in England, with limestone bands crinkled into sinuous folds by underground upheavals. Richard took chances both with his own life and—alarmingly to his wife and neighbors—with the lives of his children. It probably hadn't occurred to him that there was any other way to parent. Mary loved him for it.

During any fossil outing, Richard would have been on a mission. With brow furrowed and eyes narrowed into slits, he would have scanned the surrounding rocks as he and the children pigeon-stepped at the base of Church Cliffs in search of treasure. The alternating streaks of gray shale and lighter limestone always appeared to be glowing against a backdrop of receding dark storm clouds. Richard would have craned his neck, studying the earth for the remains of life-forms it had once sustained. The children, too, would have poked around for evidence of early beasts, gleefully collecting coiled-up worms, or ammonites, as if they were pieces of gold. Many days they would have found a few of the giant vertebrae known as verteberries as well as the lovely sea lilies. There might also have been scuttle, or primitive cuttlefish, along with some arrowheads. The threesome would have snatched up anything that looked pretty or at least remotely interesting.

There would have been countless excursions like this one, always reaping bewildering discoveries.

On a summer day two years later, in 1807, Richard set out alone for the village of Charmouth, hoping to sell some of his curios there. The London-to-Exeter stagecoaches stopped in Charmouth but avoided the steep hills down to Lyme Regis. Richard believed he could peddle his items to coach passengers when they stopped for a break outside the Pilot Boat Inn. With his sights set on good sales, he took a shortcut along Black Ven. Disoriented by the early-evening sea mist, he lost his footing, tumbling over the cliff to fall dozens of feet. Boulders the size of rowboats were knocked loose and plunged down around him. He was knocked unconscious, although only briefly. Although he also hurt his back badly, somehow he managed to stumble back home, taking to bed with a hot water bottle.

A few days later, Lyme Regis shrugged off its sea mist, serving up ample sunshine instead. But Richard remained morose. Most likely Mary begged her father to go fossil hunting, but his enthusiasm had waned.

Perhaps it was the fall, or perhaps it was the lack of medicines to help his pain, but scrounging for fossils, often in the cold, no longer held the same kind of allure for Richard. He managed to work a little on some days, but over the next few years his body was racked by terrible fits of coughing and he began to spit up blood. He died on November 5, 1810, with Mary, Joseph, and a heavily pregnant Molly at his bedside. Most likely he'd been killed by the ubiquitous consumption—later known as tuberculosis—that was the undoing of so many other Lyme residents.[17]

Richard had begun his married life with an eye toward creating something better for himself and for his family. But his goals were forever out of reach. Most of his children were dead—only Mary and Joseph, aged 11 and 14, remained—and his pregnant wife was weary. Richard had wanted excitement. He had wanted some kind of stature. In 1810, it would be up to young Mary to live out his dream.

2

A Fantastic Beast

Richard's death in 1810 at just 44, when Mary was 11 and her brother, Joseph, was 14, upended his family's whole world. Not only did it leave his pregnant wife and children virtually destitute, but all three of them—Molly, Mary, and Joseph—were so staggered by the loss that they were barely able to muddle through their daily routines. It was as though something toxic hung in the air, with neighbors smothering them with painfully sympathetic smiles everywhere they went. Molly, in particular, must have felt herself sinking beneath the £120 of debt Richard had left—possibly as a result of speculating on land or a home in 1808—a huge sum at a time when most laborers earned less than 10 shillings a week.[1]

In those days, death did not cancel debt, and if the debt was not paid, Molly could be tossed into prison. To make matters worse, the Annings did not own their home, so they had the additional burden of monthly rent payments. Molly was determined to avoid prison—and also determined to avoid the poorhouse. Under the Poor Laws dating from the Tudor times of 1485 to 1603—when there were many more poor people than rich ones—the indigent typically were housed in one of 15,000 poorhouses in England, where residents were stripped, searched, and made to feel so full of shame as to dissuade others from seeking public assistance.

Known also as parish workhouses, these cramped institutions infested with rats, lice, and roaches were run by private contractors and generally constituted a last resort, even for those who were close to starving. Until reforms were introduced in 1842, meals in these poorhouses were eaten in silence, and to add to the inhabitants' humiliation, some institutions didn't even allow cutlery. Worst of all, by entering a poorhouse, paupers generally were separated from their spouses and made to forfeit responsibility for their children, who often were forcibly apprenticed without the permission or knowledge of their parents.

Molly would have shuddered at the mere thought of it. For her, the only viable alternative was "parish relief," or welfare, and Molly applied for the funds within three weeks of Richard's death. But even this—three shillings a week—was scarcely enough to stave off hunger. Although the Annings were grateful for the stipend, which would continue for several years, Molly likely was cringing with embarrassment. For anyone coming from a respectable family, having to ask for a handout was a shameful admission of defeat. But she was out of options. At least parish relief afforded the Annings a basic diet, even if it was a monotonous one, most likely consisting of milk broth and oatmeal for breakfast and lunch, varied occasionally by pease broth—dried peas with a little mint, sugar, and sometimes pepper for flavoring—and oatmeal for supper, perhaps with a slab of bread. Luxuries the Annings had long learned to live without included meat and vegetables—but even some of Mary's better-off neighbors were rarely able to afford such delicacies.

The Annings weren't the only ones barely treading water. While the monied upper classes vacationing in Lyme Regis from London were living it up at the seaside, agricultural workers in Dorset were among the most impoverished in the country. A few of the Annings' neighbors were even starting to look at their own hair as simply another crop or resource that could be harvested and sold off for wigs and hairpieces.[2] From time to time, a special kind of "barber" was seen making the rounds in Lyme, knocking on doors to ask if there was anyone interested in getting a "haircut." If so, he'd rapaciously shave it all off and then smother the bald scalp with oil. Several months later, he'd return with the hope that, by then, the hair was a marketable length again. Selling hair was a tidy business. Not only was it valued for wigs but, until the mid-1800s, it also was desirable for its use as stuffing for tennis balls.

No one knows if the Annings ever sold their hair, but few would have blamed them if they had. The year 1810—a period of discord and scarcity for the entire country—was probably the worst time possible for a family to lose its breadwinner.

Since 1793, the war between Britain and France had raged on, an era of conflict that would last until 1815 with few interruptions. England's southern coast, including Lyme Regis, bore the brunt of the battles. The ruthless Napoleon, set on strangling Britain's economy by preventing British goods from being exported to Europe, closed off all European ports under his control, also part of his drive to weaken the British military through the disruption of trade.

Not only were port towns such as Lyme Regis hit hard by this plan to block British trade—called the Continental System—but Napoleon's hints at a cross-Channel invasion would have kept the Annings and everyone else in town constantly on edge. From the turn of the century, Napoleon's "Army of England" was perched like a tiger on the coast of France, waiting to pounce, propelling clouds of fear across the Channel. Just after Richard's death, Napoleon's ambitions started paying off.

In 1811, British exports plunged to not even one-fifth of their 1810 level. Most distressing for everyday families was a disruption in grain imports—which helped feed the nation—that sparked a serious scarcity of bread. Next, a spate of business failures and strikes played out across the country. Ever since Mary had been a small child, she would have been surrounded by people who harbored a real disdain for the French and their perceived aggression. Even in Mary's own household, there's little doubt that hostility would have been rising like a bubble during those years in the face of the continued bloodletting in Europe.

Furthering the economic turmoil in Dorset around the time of Richard's death was an embargo on exports to Britain imposed by U.S. president Thomas Jefferson in 1808 in retaliation for the British practice of impressment. Always in need of more men, British ships were known to stop American ships in order to capture sailors—sometimes violently—and force them to serve in the British navy. The final straw had been the so-called Chesapeake incident in 1807, when the USS *Chesapeake* on the high seas off Boston was approached by a British vessel, which demanded to board so that it could reclaim "deserters" working with the United States. The Americans refused. The British ship opened fire, killing and wounding several men. In the end, the outgunned *Chesapeake* had to surrender four sailors to the British.

Americans were outraged but Jefferson's cooler head prevailed, and he opted first to retaliate only with an embargo, although the United States would go on to declare war on Britain in June 1812. This embargo resulted in almost no trade between the United States and Britain until the end of the war, ostensibly by the Treaty of Ghent in 1814, although the Battle of New Orleans actually took place after the treaty was signed because news of the peace treaty was so slow to reach America. The U.S. embargo—coupled with the ongoing tit-for-tat retaliations between the French and the English—wreaked havoc on a Dorset economy already crippled by the weakening of its shipping industry and other linchpins.

Even though the source of her next meal was unsure, there was likely much more on Mary's mind than money or even her rumbling stomach: the loss of her father. After a lifetime of trailing her father like a pilot fish, his absence would have hijacked her thoughts, leaving her feeling numb. Not only was there no money, but Mary's future would have seemed devoid of fun and function as well. For years, the only escape from the drudgery of her daily chores had been her Sunday forays to the beach with her father. Time and again, Mary had been admonished by her mother, who apparently was even prone to scolding Richard like a child for tarnishing the family's good name in the eyes of upstanding neighbors. George Roberts wrote that Molly was "wont to ridicule Richard's pursuit of such things."[3] Years later, Mary would tell friends that it was her father who had always inspired her. Memories of their outings likely exacerbated her sadness. Even the air she breathed, so often dank with the threat of rain, would have been redolent of their days together. A good storm swooping up the Channel had always lit a fire under Richard, as it would break apart the cliffs and shake out the fossils, thereby doing much of the work for him. Probably not a day went by when she didn't think of her father or the shore or the whims of the weather. Taunting her was a tiny view of the water from the window above her bed. She might have closed her eyes often, imagining she was scurrying over the landslips, dangerously close to being caught between the cliff and the beach, with her father nearby, warning her to watch where she was going.

Mary's mother would have been little comfort. Most likely Molly felt completely out of her element as she took a cold, hard look at the future.[4] Adding to Molly's heartache was the death of her baby, Richard, named after his father, who was born and christened late in 1810. He followed his father to the grave in 1811, dying of unknown causes. With eight of her children and her husband ripped from her life, and with Mary and Joseph's futures squarely sitting on her shoulders, she would have scarcely known which way to turn. Often Molly had frowned on her husband's beachcombing foolhardiness, but she likely had been open in her affection for him. Despite a string of hardships, Richard had never lost his ability to draw a smile from her. "Laughter is my characteristic," he often told her, as if she had not noticed. He had always been her ballast, the vital center of their family life.

With so much weighing on her mind, Molly's two remaining children became an afterthought and were left to their own devices. Days passed, and eventually weeks, during which mother and daughter were constantly in each other's presence but spoke very little to one another. Molly even paid little heed to whether Mary was continuing with her education. But it probably didn't seem to be of much consequence anyway. In those days, most people thought that intellect in a woman was something to be avoided, not nurtured, and that educating girls, especially lower-class ones, was a waste of precious time.

When Richard was alive, beginning around her eighth birthday, Mary had been allowed to attend the Dissenters' Sunday school at the Independent Chapel. Most significantly, the school emphasized reading and writing rather than religion, and—unlike other schools—boys and girls studied together in the same classroom, a radical innovation at the time.[5] As Mary learned to read, and read well, a proud Joseph presented his little sister with one of his prized possessions: a bound volume of the *Dissenters' Theological Magazine and Review*. Books and magazines were expensive, highly prized commodities in those days, and Mary apparently held on to the volume for the rest of her life. The Annings' pastor, the Reverend James Wheaton, had published several articles in the review. Two of the essays act as a window into the religious world of Mary's childhood: One insisted that God created the universe in six days and another urged Dissenters to study a new science called geology.

Mary had always loved school but after her father's death, there might not have seemed to be much point in going. What's more, she may have been afraid to leave her mother on her own in the house for too many hours on end. Besides, with so many families dependent on charity, older children were expected to help out with any number of money-earning tasks—horse holding, running as messengers, and cleaning or other domestic work. Even children as young as seven were paid to perform errands such as bird-scaring and minding grazing animals.

Around the age of 15, about a year after Richard's death, Joseph was apprenticed to an upholsterer on Coombe Street named Mr. Hale. The job was little better than being a servant, but at least it enabled him to learn the ins and outs of the upholstery trade. Mary was left to seek odd jobs around town. The wife of a local landowner, Mrs. Stock, sometimes paid her to run errands starting around this time.

Anna Maria Pinney, a frequent visitor to Lyme Regis who would later become one of Mary's closest friends, once noted in her diary that, from Mrs. Stock's vantage point, Mary was a "spirited young person of independent character who did not much care for undue politeness or pretense." According to many historians, this was probably a fairly accurate characterization. Mary might have been bitter over the fact that she wasn't allowed to learn a proper trade like her brother. To her, it might have seemed extremely unfair. And she was the type of blunt young girl who wouldn't have been afraid to speak her mind. At the age of 11, she was already extraordinarily intelligent and articulate despite her rudimentary schooling. She seemed to be blessed with an extensive vocabulary that allowed her to easily converse with adults. No doubt she had benefited immeasurably from her father's interest and encouragement. The time spent in his company, and in that of his customers, surely fostered in her a certain level of maturity.

One evening, perhaps as she was braiding rushes for a new rug, a storm blew in, with strong winds and lashing rain that battered the windows of the timber cottage like birdshot. Later that night the storm grew worse, with waves along the shore topping 20 feet, spitting logs and all manner of debris up onto the beach. As she lay in bed, Mary would have tried hard to go to sleep, but the claps of thunder would have kept calling her back.

The next morning, curled up under heavy blankets, in a home still black and completely quiet, probably all she could think about was the seashore and how much she missed it. The prospect of spending yet another day with her maudlin mother would have been unbearable, infusing Mary with fresh determination. The long, lonely hours spent alongside her mother had only made her miss her father more. And so she arrived at the most important decision of her life in her bed that cold morning, the significance of which would become more and more apparent as time went on: She decided to return to the beach.

At first light, she would have gathered a pick, chisel, hammer, and a small gray rucksack, then tiptoed past her mother, who had taken to rising later and later each day. Whether her mother noticed the movement was hard to say. But even if she had, most likely she would have assumed it was Mary gathering the yoke and buckets to fetch water for washing up ahead of everyone else. Mary might have cast a glance at her father's old carpentry workshop, once a haven where they had rubbed and cleaned and polished so many beautiful stones together but which in the last few months had become a sobering reminder of happier times, strewn with dust and cobwebs.

The bracing early-morning air would have invigorated Mary's senses. She quickly did a check around the densely packed Cockmoile Square, most likely relieved to see no signs of life. Next door, was the Bennetts' daughter, Ann, the same age as Mary, although the two had never been close. She might have wondered if Ann was still sleeping. From the square, she would have headed east and then made a sharp left turn before passing Church Street. From the bend in the road she ambled into Long Entry, which led to the seafront. Eventually Long Entry spilled into a path that ran across Black Ven and then on to Charmouth. When Mary made that turn into Long Entry, her pace might have quickened. At this point, she would have known that she was crossing through an invisible barrier, leaving behind months of grieving fueled by Molly's relentless fatalism.

For anyone, especially a young girl, the path's footing was precarious. When Sir Stephen Glynne, a Welsh politician and landowner, visited in 1825, he noted that "Lyme is certainly not a very pleasant place of abode, from the great dirt and narrowness of the streets in the old town, as well as the inconvenient descent to the sea." With winds constantly whipping about and the thin lanes gluey with mud, every step of the journey down the disintegrating hills would have required Mary's attention.

Eventually the girl would have reached the lonely landscape she had been longing for, and the familiar sequence of limestone and shale layers of the Blue Lias in Church Cliffs would have greeted her like an old friend. On this particular morning, the cliffs might have been, as they often were, shrouded in a gray mist that—like an invisible blanket—wrapped her inside the irresistible world of the shore. Breathing in the sharp salty air, she spent the next few hours scouring the crumbling black marl, a finely textured clay containing limestone nodules that sullied her skirt and hands beyond recognition. It wouldn't have mattered. It was wonderful to be back on the beach. But, so soon after her father's death, her good time might have been marred by questions. Had all those walks on the shore with her father been too much for him? Why had she always pestered him to stay out in the cold so long? She might have recalled the way her father used to affectionately refer to her as "Mary girl" while they stalked the unusual stones together, side by side, and her eyes would have begun to swim.

Fortunately, the powerful storm the night before had afforded a bonanza of finds to distract Mary from her memories. But after hours of perusing the pebbles, hunched over like an old woman, her shoulders and back would have begun to ache. It was time to go home. But first she wanted to take a last stab at a particularly massive mound of clay. To her delight, she heard the sharp telltale clank of a fossil-bearing nodule after hitting a clump of limestone with a hammer. Using her small pickax, she carefully chipped away at the hard surface and was, after a good 30 minutes, rewarded with a fairly striking ammonite, or snakestone.

By now, it was approaching late morning, and possibly she was growing hungry. Mary hurriedly laid out the ammonite and her other treasures so that she could pick out the ones worth saving. Although weary from a morning of hunting, she was pleased with her haul. Caught up in her own thoughts, she hadn't even realized that there was another person walking on the shore that morning, let alone that this person was now trying to engage her in conversation. Historical records differ as to exactly what happened, but she might have stood up, startled. It would have been rare to see anyone on the beach this time of year. But before her was a lady, one who was too nicely groomed to be out beachcombing. Mary didn't recognize her. Perhaps she was a visitor to Charmouth.

Mary probably asked what she wanted, and the lady replied that she was interested in buying her curiosity—the ammonite. The lady said she'd not seen another one like it. Most likely, Mary was so dumbfounded she said nothing. The lady apparently mistook Mary's hesitation for a reluctance to part with the curiosity. "I'll give you a half crown for it, little girl. I always like to take home an unusual stone from each beach that I walk on during my holidays, and I have seen nothing as beautiful as the one you have found," the lady might have said. Prior to decimalization in 1971, Britain relied on a monetary system consisting of pounds, shillings, and pence. The smallest unit of currency was a penny, the plural of which was pence, with 12 pence per shilling and 20 shillings per pound. Five shillings equaled a crown; two shillings and six pence equaled half a crown.

As if in a dream, Mary passed her the ammonite and took the money in return. In her hand she held a whole half crown![6] And it was for a fairly ordinary curiosity at that. The lady may have wondered if Mary knew anything about the particular type of stone. Mary would have known only what her father and others had told her, that ammonites were revered by lots of those in Lyme for their alleged magical powers. Some people believed they were safeguards against snakebites while others believed they were a cure for blindness, impotence, and barrenness. Still others thought they were helpful for insomniacs; if you placed one under your pillow, it would somehow lull you to sleep. The very superstitious even painted a snake's head on ammonites and then wore them as protective charms. But Mary probably said nothing of these things to the lady. She simply shrugged her shoulders and prayed to herself as hard as she could that the woman wouldn't change her mind. For Mary, a whole half crown would have been equivalent to tremendous wealth, enough to buy some bread, bacon, and maybe even some tea and sugar, enough to last a week. From that moment on she was fully determined to go down "upon beach" again. After muttering a brief thank you, Mary gathered up the rest of her fossils and ran all the way home.

When she burst through the door, so full of life, her mother probably thought Mary had gone mad, such was the contradiction from her demeanor during the previous days and weeks. But then Mary showed her mother the half crown and spilled the story of the sale. A half crown was more than anyone had ever paid Richard for an ammonite. It was a wondrous sum. But Molly might have seemed incapable of accepting their good fortune, at least initially. As if snapping out of a deep sleep, Molly likely peppered her daughter with questions: What if the tide had rushed in and swept you away? What if you had had to make a fast getaway over the cliffs, cliffs so perpendicular they rose up like walls? What if a falling rock had pummeled your head? Any mother would have dwelled initially on her child's safety, although in this case it was likely the first time in weeks Molly had paid heed to her parental responsibilities. But despite her concerns, after learning all the details, even Molly wasn't able to deny that selling curiosities may be their last best hope for survival. And even if her mother had forbidden her from collecting curios, it probably wouldn't have made much of a difference to a girl with as formidable a personality as Mary's.

Weeks passed, and the mud-splattered figures of Mary and Joseph lurching over landslips, bowed down beneath the weight of their heavy gray rucksacks, once again became a familiar sight on the beach. But initially they failed to come across any truly unconventional finds. Even so, for the pair, the pleasures of being together again, at the edge of the sea, would have far outweighed any disappointments. Both likely were revived and chirpy and in sync with each other, just as they used to be when their father was alive.

They remembered well how splitting open the cliffs' limestone nodules had always been a bit like playing the three-shells game, in which someone gathers together three shells, places a small item under one, and then quickly shuffles all the shells around so that it's impossible to know where the item has landed. Sometimes there might be a pristine ammonite inside a nodule, maybe even something better, but most of the time they turned out to be empty. Often what looked to be the most promising kind of stone turned out to be about as fossiliferous as an average house brick when split open. But the picture of brother and sister, heads together at the table in Richard's old workshop, engrossed in sorting through their daily takings, would have brought an unfamiliar smile to Molly's face.

Mary and Joseph rarely allowed themselves to become disheartened. After years of watching their father's efforts, they would have known that a good fossil hunter was a patient one. Day after day, the two plodded on. It was not only interminable work, but lonely work as well, with the two often separated by large stretches of beach, with seagulls and curlews flying overhead, their cries often the only sign of life around. That was the nature of their toils. Fortunately, they were always able to uncover an item or two worth peddling to tourists.

Once a week, like clockwork, as a coach pulled up to the Three Cups Inn less than a block from the Annings' home, and unloaded visitors from Bristol, London, and other cities, Mary and Joseph became a reliable spectacle. Always they were there, beside their father's round table, which by now was covered with a mess of bones, rocks, and shells. Mary lacked any kind of sartorial style and indeed was an eyesore with her tangled dark hair, dowdy muslin skirt, and worn clogs. Yet her business acumen was remarkable for someone her age, and she was able to hold her own with any of the "better" passersby. By this time, even more families of solid means were making annual treks to Lyme Regis for the "season," which ran from May to October. They eagerly bought curiosities with whimsical names like cupid wings, ladies' fingers, and devil toenails. Even so, there would not to be another sale as large as the one of the ammonite, at least not for a while.

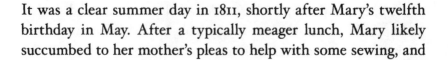

It was a clear summer day in 1811, shortly after Mary's twelfth birthday in May. After a typically meager lunch, Mary likely succumbed to her mother's pleas to help with some sewing, and an antsy Joseph decided to go rooting around the beach on his own. Among the exposed fallen rocks, something dark glistened, an outline barely visible to the casual eye. It had a strange profile, hardly conspicuous in the foreshore just below Black Ven. Joseph gingerly unearthed as much of the object as he could from the malleable clump of shale that surrounded it. Eventually what emerged was the giant head of some kind of creature, perhaps the skull of an enormous lizard. About four feet in length, the skull had room for huge eyes and at least 200 teeth.

Joseph surely thought it must be the bony framework of a crocodile's head, if not a lizard. He ran back to town, imploring some local men with a soft spot for the industrious lad to help him dig it out and carry it back to his father's old workshop. At first glance, they too thought it was a bewildering find. And certainly it was massive, whatever it was. But, in truth, a crocodile skull—and certainly Joseph saw no reason to believe it was not such a skull—wasn't all that special. It certainly wasn't noteworthy enough to generate much more hoopla than a really lovely snakestone. After all, the children had found plenty of what they were sure were crocodile teeth in the past few weeks. This, Joseph believed, was simply the head. But Mary might have had her doubts.

Joseph was pleased with what he'd found but wasn't able to give it a great deal of thought in the coming weeks. Around this time, he began funneling more and more energy into his apprenticeship; he didn't have as much time for fossil hunting. After all, the money from the sale of Mary's ammonite had long run out, and the family had to eat. And so Joseph—now the head of the household—asked Mary to search for the rest of the skeleton, if and when she was so inclined.

Mary wouldn't have needed to be asked twice.

The place where Joseph found the fossilized head was perhaps more than a mile from their house, near Charmouth, and the rest of the beast could have been tucked away anywhere in the minimountains of rough fallen boulders. Indeed, Mary searched for nearly a year before another terrible storm blew through and placermined the cliffs in such a way that the first few bits of a skeleton were exposed, locked in the cliff several feet above the shore. Surely these were the bones she had been searching for, Mary eagerly began excavation. Working with her hammer, chipping away at the rock, initially she found several large vertebrae, up to three inches wide. As she picked away, little by little, it was possible to make out a few ribs embedded in the limestone, with several still connected to the vertebrae. She managed to convince some of the men in town to help her extract these first few fossils from the cliff side. Gradually, they revealed an entire backbone, made up of 60 vertebrae.[7]

On one side, the contour of the skeleton was plain enough to make out; it was not unlike a largish fish with a lengthy tail. On the other side, though, the ribs were harder to distinguish, pressed down upon the vertebrae in such a fashion that it looked to be little but a messy mass of bits and pieces. But, eventually, and bit by bit, Mary was able to draw out the bizarre creature from its ancient tomb. It became apparent to those who saw it that this was no simple crocodile but some other giant animal, one up to 17 feet long. It took months, but eventually Mary found almost all the bones of the colossal creature. Molly, too, got in on the act, pleading with some of the men at the quarry to help her daughter by carefully chipping away at the biggest and heavier chunks of surrounding rock, without damaging the skeleton.

By this time, months into the excavation, most everyone in town was aware that Mary had found something truly unbelievable in the cliffs. Analyzing her discovery, Mary herself felt sure she was tracing the outline of what must be some kind of monster. When it was finally completely uncovered, it took several men to raise the heavy skeleton off the cliff side—where it had lain untouched for 175 million years—and to lug it all the way to Richard's shop. In every way, it was fantastic. The creature had flippers like a dolphin, a mouth like a crocodile, and a pointed snout like a swordfish. Its backbone was like the spine of a fish, but the chest could have belonged to a lizard.

Mary knew exactly what this meant: The skeleton was a much greater discovery than the skull had originally indicated. The realization hit both Mary and Joseph like one of their hammers.

Eventually news spread far and wide that a young girl from Lyme Regis had made an incredible find: an entire connected skeleton of a creature never before seen.

In no time, Mary's *Ichthyosaurus,* as it was eventually named in 1817, was snatched up for £23—enough to feed the family for well over six months and possibly even to help start paying off Richard's debt—by the lord of the manor, Henry Hoste Henley, who lived less than a mile away at Colway Manor. Henley, a collector for a private museum, donated the specimen to William Bullock's Museum of Natural Curiosities, or "London Museum," at 22 Piccadilly, where it was exhibited in the newly built Egyptian Hall—the closest place resembling a natural history museum at the time.[8] Henley not only owned the land above the cliff where Mary found the skeleton, but he owned nearly all the land around Lyme Regis, and so Mary probably would have had no other option than to sell it to him when he asked to buy it. Even so, £23 was a huge sum to a family like the Annings, and they had no way of knowing that the fossil would later resell for twice that much.

After the sale, Mary threw herself into fossil hunting with greater zeal than ever before, and she was encouraged in her pursuits by a new acquaintance. Tall, dark, and exceedingly droll for someone so young, Henry De la Beche breezed into town in 1812, around the same time Mary made her discovery, a 16-year-old of independent means. Mary likely learned quickly by asking around that he had inherited a large sugar plantation in Jamaica that had prospered greatly in recent years thanks to the slave trade. His father, Thomas, an officer in the army, had died when Henry was quite young.

The gregarious Henry arrived in Lyme quite by chance; he was very much in disgrace, having just been dismissed from a prestigious military college for insubordination.[9] School records show that he had entered the Royal Military College at Marlow in 1809 but then suddenly "left it" in 1811. Further reports revealed that his leaving had something to do with behavior that was well short of "the utmost deference and respect, which should be observed equally towards the civil professors as towards the military officers." Whatever the precise reason for his leaving school, he came to Lyme Regis because his mother had recently married for the third time—in one local paper, *The Lymiad,* she was rather unkindly nicknamed Madame Trois-Maris. Her new husband was a prominent businessman named William Aveline, who lived on Broad Street, the main commercial district in town, only a short walk from the Anning cottage.

Henry was a curious teenager who seemed to have natural science in his blood. Both Mary and Thomas Carpenter, the Lyme doctor and coroner who sometimes acted as the town's mayor—and who was also an early collector of fossils—encouraged his interest in rocks and shells. Eventually Henry decided to study geology, a science just taking off in popularity, but still a rare choice for a rich young gentleman.

Mary and Henry, and sometimes Joseph, spent hours scouring the wind-whipped coastline, chatting amiably. Henry was impressed at the way in which this young girl paid little attention to the waters of the ocean as they surged toward her. Generally well groomed, Henry probably couldn't help but notice her hands, rough and ruddy from constantly digging into the crevices of the cliffs. Thanks to the ever-present mist, Mary's face was raw and her hair was matted. By the end of a day, they both would have been barely able to feel their legs, so long would they have spent crouching over rock pools. But neither would have minded because they were always focused on one thing: making a wonderful find.

3

An Unimaginable World

*I*n 1812, war broke out between the United States and Britain. America invaded Canada, and its soldiers were repulsed by the outnumbered defenders in what is now Ontario, and Louisiana was admitted as the eighteenth U.S. state. The English writer Charles Dickens was born, and Sacagawea, the Shoshone woman who blazed the trails for Meriwether Lewis and William Clark, died.

In Lyme Regis, a local newspaper, the *Dorchester and Sherborne Journal,* reported on a wide range of breaking news, including shipwrecks, smuggling incidents, and accidents resulting from excessive drinking. The newspaper also devoted a small square of space to one discovery of a "crocodile" skeleton in the Dorset cliffs.[1] To the journalist, it seems, Mary's find was a commendable curiosity but not a whole lot more, uncovered in the blandness of the beach's Blue Lias layers.

But Mary's discovery of the first complete skeleton of a completely unique and grisly 17-foot-long creature with an elongated 4-foot skull—or at least the first known to scientists—was nothing short of a small miracle. In the years after it was found, the paradoxes surrounding the beast were endless. First there were the slender vertebrae, analogous to a fish's backbone. Then there were the four finlike limbs and the large shark-like tail. Other features, though, were similar to those of a dolphin or other aquatic mammals. The alien combination made it impossible to categorize the creature. Even a child with the most vivid of imaginations would have been hard-pressed to conjure up a specimen more mystifying. Looming most large, though, was the fact that the skeleton posed one of the first serious challenges to the foundation of Christianity on which society was built. No one in the world could recall seeing such a creature before. And so, in time, there arose in people's minds the question that most were probably afraid to ask: How could someone have found the remains of a creature that no longer existed when every single being in the world was designed at the same time and with a specific purpose by a loving and all-powerful God? In the nineteenth century, it was widely believed that animals did not become extinct; any unusual curiosities were explained away as being from animals living undetected in a far-off region of the world. But the puzzling attributes of Mary's fossil struck a blow at this belief and eventually helped pave the way for a real understanding of life before the age of humans.

Certainly by this time, the early 1800s, people were used to finding exotic fossils; oddities such as arrowheads, snakestones, and verteberries were standard fare, for those hunting both to the east and to the west of Lyme Regis. The Georgians who lived between 1714 and 1830, and the Victorians thereafter, were fanatical about their seashore trinkets, displaying them in lavish glass cases or using tiny shells to decorate frames, mirrors, and jewelry boxes. But the significance of these pretty, seemingly innocuous collectibles was almost entirely lost on most of those who amassed them. For some time, Mary and the others underestimated the kinds of stories the cliffs and the fossils inside them were able to tell about the history of Earth.

For centuries and centuries, Christians had been convinced that Genesis told the true story of creation. Fossils only reinforced the biblical account. For example, some fossils were found at such high altitudes that people thought they must surely have been deposited there as a result of the worldwide flood depicted in Genesis. Even during Mary's time, it was inconceivable that a completely different world might have existed here, on this planet, before humans became a part of it. After all, the Bible stated that God created the heavens and the earth and every living thing in it in just six days. There was never any mention of a prehistory and therefore never any mention of prehistoric animals. What the Bible had described, in vivid detail, was a flood so potent it wiped out everything on Earth except the animals marshaled onto the ark by Noah, two by two. In general, very few people doubted the Bible's veracity.

But beginning in the early 1800s, a process of discovery began, thanks mostly to an influential new circle of educated fossilists, linked by their skepticism of Bible-based truths, who later set the stage for Charles Darwin's findings. They were asking the questions no one else wanted to ask in an attempt to explain matters that others were happy to leave ambiguous. Where were the fossilized remains of the human victims of Genesis's massive flood? Why were so many different kinds of fossils, the remnants of different kinds of creatures, buried so deep inside Earth? Could it be possible that there was more than one creation? As time went on, they wondered: Why were the fossils of fish, buried inside one rock, covered up by so many other layers of rock that contained only land animals? These early geologists, all so inquisitive, formed an intimate clique.

Providing an excellent forum for these curious fossilists was the Geological Society of London, which was just beginning, having been inaugurated at a dinner at the Freemasons Tavern in 1807 by geologist George Bellas Greenough. Among its founding members was Dr. James Parkinson, who became known for identifying the condition known as Parkinson's disease. Not only was he a respected surgeon, but he was also a noted geologist who published "Organic Remains of a Former World" in 1804, a paper that has been touted as the first serious attempt to provide a scientific explanation for fossils. By the time of Mary's discovery, the Geological Society was on its way to becoming a highly influential and cohesive body, with members often dining together at each other's homes, obsessively debating issues late into the night. But it wasn't about to admit women, not even as a member's guest. In those days, only men with a certain amount of wealth or status could vote, attend university, or hold public office. As she grew older, Mary was perfectly placed to become acquainted with this group but, as a woman, she would only ever be an outsider.

Mary's world remained a sheltered one, where people probably exchanged views about little else but the war and whether it was ever going to end and, if it did, whether this was ever going to lead to a decrease in food prices. A handful of people in Lyme Regis wondered if fossils might be something more than just oddities flushed out of their storage spaces. One of them, Elizabeth Philpot, was establishing herself as a woman of action, able to cut through the confines of Britain's class system in order to build a strong friendship with someone young enough to be her daughter. Time and again, it was Elizabeth—probably considered upper middle class as a result of her family's money—who exhorted the lower-class Mary to think seriously about studying science and geology.

In London, the public's response to Mary's specimen was explosive. Day after day people flocked to the gaudy new Egyptian Hall, streaming in single file past the creepy frame of a creature that had been found in the remote countryside, far from London. Some came from cities far from the capital, arriving by stagecoach on improved roads that allowed them to travel at speeds of up to 12 miles per hour. But no matter how long the journey took, likely no one regretted making the trek. Visitors gaped in wonder at the round eye sockets that seemed to stare back at them menacingly when they walked passed. The skeleton was displayed alongside another crowd pleaser: the curiosities brought back from the South Seas by Captain James Cook, superb pieces including stone weapons and feathered objects. The public's fascination with the budding science of geology was in full bloom at this stage—with shell shops becoming all the rage—so it was no surprise that Cook's curiosities coupled with Mary's bewitching bones attracted a crush of visitors.

Adding to the allure of Mary's skeleton was the controversy it was causing among some of the best minds of Britain.

Sir Everard Home, who held the distinguished position of Surgeon to the King, was the first person to closely inspect Mary's specimen.[2] At the time, he was considered to be Britain's leading anatomist. But, like so many others, he was stumped. At first he claimed the find was a new class of crocodile. Then he reasoned it was a fish, though by no means "wholly a fish." Later he surmised it was a gigantic aquatic bird since the bones of the eye, he wrote, were "subdivided into thirteen plates, which is only met in birds." But if it was a bird, then why didn't it have any wings? A long five years later he was still tussling with adjectives in an effort to explain the fossil, and he changed his mind again, concluding once and for all that it was surely some kind of amphibian, most likely a cross between a salamander and a lizard.

For Mary, though, this tortured process wasn't the real travesty. In an address to scientists during which he first described the skeleton, Home apparently never mentioned her when thanking those who brought the fossil to the world's attention. He also incorrectly, and irritatingly, praised Bullock's museum for Mary's scrupulous cleaning of the fossil. Mary most likely learned of this slight later from one of the scientists who visited Lyme Regis— and it was probably one she never forgot.

If no one was able to make sense of the bones, it stands to reason that coming up with a name for the creature was going to prove to be another exercise in frustration. Indeed, the scientific community's wrangling over the name went on one excruciating year after another, finally coming to a head in 1817, after Bullock sold the specimen to the British Museum for £48 pounds, or more than twice what Mary had received for it less than five years earlier.

Founded in 1753, when the physician Sir Hans Sloane bequeathed his vast collection of plants, animals, and antiquities to the British people, the British Museum already was an important institution, a product of Enlightenment thinking built around the desire to collect and study objects from every known culture.

After a long perusal, the museum's curator and keeper of natural history, Charles Konig, simply repeated the obvious, stating that Mary's skeleton looked like both a fish and a lizard. And so, simply putting the Greek words for "fish" and "lizard" together, he named the creature *ichthyosaur*, or "fish lizard."[3] Even though the description turned out to be a misnomer, since the creature was neither a fish nor a lizard but rather a sea reptile contemporaneous with dinosaurs, the name has stuck to this day.

The puzzle of Mary's specimen weighed on the public's mind, and religious leaders knew a threat when they saw one. Convinced that Mary's ichthyosaur—as well as other bits of bone and rock newly drawn out of their hiding places—was soiling the sacred teachings of the Bible, they were becoming increasingly alarmed by the work of the early fossilists. "Was ever the word of God laid so deplorably prostrate at the feet of an infant and precocious science!" exclaimed an exasperated evangelical Anglican pastor named George Bugg, author of *Scriptural Geology*, written between 1826 and 1827.[4]

Young Mary wasn't likely to realize it at the time, but change was coming—and partly because of her. The Industrial Revolution was whipping up real power and speed—the power and speed of the machine—resulting in the kind of technical and economic progress that had never been seen before. And fossils were causing a real disruption in theological thinking, threatening to rip holes through the biblical account of creation. The church was fearful of the changing tide and wasn't about to make a quiet retreat.

In the same way that Christian fundamentalists hold sway in some American school districts in the twenty-first century, the

Anglican church in Britain in the early 1800s was a force to be reckoned with. In those days, most people attended church services every Sunday, often more than once, and most people took the Old Testament's teachings to heart. In particular, most people considered unassailable the famous conclusions published in 1650 by the Archbishop of Armagh, James Ussher, that the world was created at 8 P.M. on October 23, 4004 BC. Before 4004 BC, he had pronounced, there was nothing. This was no trivial proposition. The highly regarded scholar's conclusion was based on 907 pages of meticulously laid out chronology, backed up by dates mentioned in the Bible, including every single "begat" in the Book of Genesis. Most people—not just those in Britain—held Ussher up as a luminary. Indeed, his unequivocal analyses were appended to many a standard Bible in all parts of the world. Even as late as 1815, the biblical scholar George Cumberland put into words the skepticism of geology felt by many Christians in a letter to the editor of the popular *Monthly Magazine* when he said: "We want no better guide than Moses." But fossils were lodging a good fight against the Bible's authority, whether the church liked it or not. Indeed people were finally beginning to believe that fossils also might be keys to unlocking the mysteries of the past.

Helping to probe fossils from their hiding places was the Industrial Revolution which—thanks to a population explosion brought on by a decreasing death rate and increased fertility—led to the need for a rapid transportation system for both goods and people. Railways were proliferating; quarrying for limestone in order to make cement was expanding. Added to this was the rapidly growing mining industry. The result of this frenzy of activity was digging, lots of digging. And all of this digging was generating a prodigious number of fossil finds. The mass excavations also were bringing to light Earth's many layers of rock, one on top of the other. On closer scrutiny, it became apparent that these layers actually were a reliable record of time. Most rock is made gradually from either mud that settles and hardens or from lava that flows from a volcano. Because each layer of mud or lava gets deposited on top of older layers, rock strata are an easy-to-read chronicle of Earth's history. As one moves farther down, the rocks get older; as one moves up, they get younger.

This was easy enough for most people to understand. But what wasn't logical was the extreme thickness and depth of the rock record—many miles of layers on top of layers on top of layers—presenting evidence of a world that was likely to have existed much longer than a few thousand years. The fossils themselves didn't seem to make any sense either. In the more recent layers, the fossilized creatures looked similar to modern species, but those nearer the bottom looked nothing like anything anyone had ever seen.

One reason the early geologists were finally able to start connecting the dots was the pioneering work of an obscure but single-minded canal surveyor named William Smith.[5] Like Mary, Smith had had little formal education, but six years supervising the digging of the Somerset Canal in southwestern England in the mid-1790s had taught him a thing or two about rocks. The son of a farmer, who died when William was eight, Smith grew up collecting the "pundibs" that were strewn across his family's property. These pundibs, which made perfect marbles for Smith and his friends, were really ancient brachiopods, or marine invertebrates.

As a grown man, Smith became obsessed with the disguised underbelly of England. For instance, he noticed that each different kind of rock in a stack boasted its own particular kind of fossils—fossils not evident in any other layer. One layer of rock exposed a certain category of fossils and thereby a certain point in time. In particular, Smith was struck by how fossils in the layers were so often arranged in the same predictable pattern from top to bottom. Again and again, he came across the same sequences of fossils, even as he crisscrossed England on his meager salary. To Smith, it seemed that each type of animal must have lived during a particular span of time, spans that overlapped into those of other animals. Soon he was able to go to any part of England and describe—without looking—the order in which rocks had been formed.

But Smith's biggest accomplishment was the composition of the first geological map of England, which appeared in 1815. The intricate map, beautifully painted by hand, was more than eight feet tall and nearly as wide. Yet it had taken Smith 16 years to get it published. Never a natural self-promoter, Smith found it hard to gain recognition for his groundbreaking geological work, especially among the upper classes. Even four years after the map's publication, he still hadn't been adequately compensated—or acknowledged. With his young wife spiraling toward insanity—and rumored to be a nymphomaniac—Smith ultimately found himself bankrupt and in a debtors' prison after his maps were copied and sold without his permission for lower prices than he was asking. Eventually he fled London for the north of England, where he roamed for 10 long years in search of employment. It would be several years more before his accomplishments were rediscovered by the scientific establishment. Surviving financial hardship and the prejudices of the aristocrats, he was finally recognized as a pioneer in 1831, when he received the Geological Society's highest award, the Wollaston Medal.

Today Smith is considered to be among the first to recognize that fossils are really time markers in rock formations. And his map was the first to illustrate hundreds of millions of years of changing landscapes.

To say that the questions raised by fossils were disconcerting was an understatement. Many of the early geologists were themselves men of God—indeed, many were paid handsomely out of church coffers. As such, they tried desperately, through a variety of theory twists and other mechanisms, to reconcile geological findings with Christian teachings. Some thought surely God had reacted to various catastrophes with successive creations. Others reasoned that some latitude had to be given in the interpretation of the Bible's wording. Certainly most churchgoers still believed that God literally created the whole world in six days, but others were starting to see the six days as an allegory for a much longer period of time.

Eventually, the issue split early geologists into a handful of contentious camps. Around the beginning of the nineteenth century, the charismatic German mineralogist Abraham Werner boldly asserted that Earth had started as an ocean of water, which later precipitated the solid rocks that today form most of the dry land.[6] His theory, called neptunism, regarded Earth as a static entity—at least until the next divine catastrophe came along to transform it. But Werner had his fair share of critics, the loudest being Scottish natural historian James Hutton. Hutton argued that Earth's creation was a theological riddle and that geologists had no business trying to solve it.[7] He believed the planet was extremely old, with modern geological structures springing up slowly due to erosion and other processes that still can be observed today. He famously described Earth as having "no vestige of a beginning, no prospects of an end." This, and other related ideas, came to be known as uniformitarianism.

These two theories, however, were at odds with the findings of still another big thinker, Swiss natural philosopher Jean Andre De Luc, who was a staunch defender of Genesis. He concluded, leaving no room for debate, that the six days presented six epochs that ended with the flood.

Over time, each viewpoint garnered its own vociferous champions who became embroiled in intellectual and sometimes even physical feuds over the repercussions of studying fossils. All the while, the public looked on with fascination and unease. Indeed, so much suspicion was being drummed up by the studying of rocks and fossils that a sizable portion of the population referred to it as an "underground science"—or a subversive activity.

The most thorough and enlightened insight into the fossils' secrets was coming not from London but from Paris, where an inspiring French naturalist named Georges Cuvier made waves during a 1796 lecture with a mind-blowing assertion: Some species have forever vanished from the face of Earth.[8] It wasn't evolution he believed in, but the idea that each species originated independently and remained unaltered until it became extinct. At the time, merely suggesting such an idea was considered outrageous and even downright blasphemous.

Again, some scientists tried to counter Cuvier's arguments by interpreting the fossils as the remains of living species: Fossils of mammoths found in Italy, for example, were called the remains of the elephants brought by Hannibal when he invaded Rome. But Cuvier was one of the most persuasive orators in history, and he could inveigle even the most stubborn among his critics to consider his point of view.

Born in 1769 into an educated family in the French-German duchy of Wurttemberg, Cuvier was a whip-smart child pushed to achieve academically by a hard-driving mother. His formal education included a solid introduction to natural history, a broad subject that included biology, geology, and paleontology.

As a young man, Cuvier served as a private tutor for a noble family in Normandy, where he was able to witness the French Revolution from a comfortable distance. After becoming a citizen of France in 1793, when the French government annexed his homeland, Cuvier accepted a post in the revolutionary administration of Normandy. Vehemently opposed to the ferociousness of the central regime's Reign of Terror, he poured all his energy into zoological fieldwork. In 1795, after a moderate republican government took power in Paris and promised to rebuild the core scientific establishment destroyed by the Terror's handiwork, Cuvier moved to the capital in search of a career in science.

There were plenty of opportunities for a naturalist of his obvious brilliance and ambition. In no time, Cuvier gained an assistantship at the renowned Muséum National d'Histoire Naturelle. With his arresting shock of wild red hair and dazzling blue eyes, he made a striking impression, and it wasn't long before the charismatic naturalist had earned a reputation to match. His subsequent rise was meteoric, and the study of natural history would never be the same. Soon Cuvier focused his considerable talents on the field of comparative anatomy.

Cuvier was helped in his pursuits by having at his fingertips in the museum a remarkably complete collection of the world's mammals—past and present—which allowed him to make definitive distinctions among them. He made the most of his advantage, sharing the material only with his collaborators and protégés.

But, again it was in 1796, three years before Mary was born, that the 27-year-old Cuvier truly stormed into the spotlight in France after presenting his controversial paper on living and fossilized elephants to the Institut de France. Beyond establishing that African and Indian elephants were different species, it also put forth the wild notion that mammoths were actually a separate species from any living elephant and therefore must be extinct. Thus Cuvier introduced—for the first time—the fact of extinction, one of the foundations of paleontology.

Cuvier's evidence and his presentations were mesmerizing. Considered by many to be the greatest intellectual mind in Europe at the time, he beguiled audiences with animated and highly illustrative comparisons of the skeletons, showing, for example, that mastodons and mammoths were different from elephants in that they did not have ridged teeth. More convincingly, he argued how unlikely it was that these animals—8 to 10 feet tall, weighing up to six tons—were still roaming the world without anyone noticing them.

Cuvier believed that Earth was immensely old and that, for most of its history, conditions had been more or less similar to those of the present. However, he believed that periodic catastrophes had befallen the planet and that each one had wiped out a number of species. Cuvier regarded these catastrophes as events with natural causes, causes that remained an important geological mystery. Although he was a lifelong Protestant, Cuvier never explicitly related any of these catastrophes to biblical or historical events. However, other early geologists suggested that the most recent catastrophe must have been the biblical flood.

One person closely watching Cuvier's work was U.S. president Thomas Jefferson. A longtime fossil enthusiast and also a devout churchgoer, Jefferson felt certain that the giants described by Cuvier were hiding somewhere in the vast wilderness of the American West. He even implored Lewis and Clark to seek out the creatures during their trek in 1803 and was certain they would return with tidings of mastodons, dead or alive. By this time, a debate over extinction also was stirring in America, where huge bones and teeth, weathered out of farmers' fields, were initially described as belonging to giant animals drowned in the biblical flood.

In 1801, American artist, naturalist, and museum proprietor Charles Willson Peale excavated some of the biggest bones yet from a farm in the Hudson River Valley, bones that belonged to an unknown species, later identified as the mastodon.[9] The 11-foot-tall skeleton became the star attraction at Peale's Philadelphia Museum, which had begun modestly with only a few stuffed birds but soon grew to an institution held in high esteem.

To Peale and his friend Jefferson, however, the mastodon was far more than an exciting exhibit; it was a natural rallying cry for a new nation. Prior to the American Revolution, the great French natural historian George Louis LeClerc de Buffon had proposed that the environment in North America was so impoverished compared to that of the Old World that it could support only a weak and degenerate fauna. Buffon and his followers insisted that when subjected to North American conditions, even the largest and most robust animals would shrink in stature and wither in vigor. The discovery of the mastodon actually helped redefine America's image as a more powerful nation.

But the failure of the Lewis and Clark expedition to come across any sign of these monsters shook Jefferson's core beliefs about nature and forced him to concede that perhaps extinction wasn't such a crazy notion after all—and that it could quite possibly be a scientific advance worthy of further investigation.

But extinction continued to be an unsettling, unthinkable concept. Like many prominent eighteenth-century thinkers, Jefferson had believed since childhood in a perfect creation under divine supervision, one in which no new kind of animal could come into being and no living species could be destroyed. "Such is the economy of nature," he wrote in his *Notes on the State of Virginia* in 1781, "that no instance can be produced of her having permitted any one race of her animals to become extinct; of her having formed any link in her great work so weak as to be broken."[10]

Cuvier's conclusions were destined to make bigger inroads among a more receptive audience in France, where the church held less of a stranglehold on intellectual progress than in America and Britain. Ironically, Cuvier's most compelling arguments were based on studies of fossil mastodons, mammoths, and sloths from America—specimens that in some cases Jefferson himself had supplied. These studies convinced many that extinction had become a well-established theory. Indeed, studies were showing that mastodons—cousins of the woolly mammoth—had been extinct for some 10,000 to 12,000 years.

At the same time, one other French intellectual in the early 1800s was deriving a far different conclusion from the examination of fossils—Jean-Baptiste Lamarck, a botanist and invertebrate zoologist.[11] From his studies, he determined that some species had remained the same over millennia while others had changed. Lamarck, born in 1744, proposed the first theory of evolution, nearly a century before Darwin published his *Origin of Species*. His ideas are outlined in his book *Philosophie Zoologique,* published in 1809. His theory was called transmutation; the word *evolution* did not come into everyday usage until well after *Origin* was published in 1859. (Evolution is from the Latin "evolvere," which means to roll out.) Briefly, Lamarck differed from Cuvier in that he was confident organisms could transform in such a way that higher forms could emerge from lower ones. In his view, the giraffe's ancestors had stretched their necks in search of food and then passed this "acquired characteristic" on to their offspring.

Taking the theory a step further, a blacksmith who develops strong arm muscles will pass his biceps on to his children. What is now called Lamarckism was proved false long ago by genetic theory, which explains that a giraffe's long neck derives from a genetic structure that contains the DNA for the body to produce a long neck—and not from stretching high up to reach food. But Lamarck's assertion that if you don't use it, you lose it was viciously maligned even during his lifetime.

Overall, Lamarck's evolutionary theories raised hackles not only for their religious implications but also because of fears that the lower classes might take them up as a battle cry; if life could progress and improve, why couldn't they? Beyond inciting apprehension among the upper crust, Lamarck also had the very bad luck of attracting the disdain of Cuvier, who wrote a scathing eulogy when Lamarck died. Although Lamarck's theory was an ingenious interpretation, Cuvier seems to have done all he could do to undermine Lamarck and his work. At the same time, Lamarck's criticism of the highly revered Cuvier and his anti-evolutionary stance lost him nearly every one of his friends. Lamarck died blind, distraught, and alone except for his devoted daughters. Poverty even chased him out of his grave. Five years after his burial, his rented plot was sold to someone else, and his remains were removed and dispersed to an unknown location.

Back in Lyme Regis, Mary knew little of this vast pool of knowledge being assembled in dribs and drabs and debated across Europe and America. In Mary's world, there were few books to open, few discussions to engage in, no lectures to attend, and not even one museum to visit. She was likely desperate to learn about the wider world, but her humble status and her family's penury left no hope for a formal education or even access to books. Sometime in the early years of the nineteenth century, Lyme Regis opened two small libraries that loaned out books for a small fee, but few had anything to do with science. Mary knew no other life and probably wasn't able to imagine one. Her knowledge had been built by the seashore and her father's smattering of facts, as well as by the limited tutelage of childhood friends such as De la Beche and Philpot. If she had any chance of attaining real and sustained intellectual stimulation, she probably didn't see it. But all that was about to change.

Improved roads and the introduction of carriages meant greater mobility during the early 1800s that was soon matched by the greater dissemination of information. It wasn't long before news of Mary's skeleton reached all corners of academia. And when it did, no one was more intoxicated by it than the oddball Reverend William Buckland, an ordained minister and one of the early pioneers of the new field of geology who was to be a lifelong friend to Mary.

Born in 1784, Buckland had grown up near Lyme Regis, in the quaint village of Axminster, just six miles inland from the Dorset coast. A love of geology ran in his family. Buckland's father, the Reverend Charles Buckland, took his son on long walks in the countryside where an interest in road improvements led to a fascination with rocks, shells, and fossils. In contrast to the Anning household, the Buckland household was an intellectual and rambunctious one in which books were always open and theology was always debated. Finances were of little concern. But like Richard Anning, Charles Buckland fed his child's curiosity about natural history by taking him to explore local rock quarries and even the cliffs close to Lyme Regis. "They were my geological school," Buckland later wrote. "They stared me in the face, they wooed me and caressed me, saying at every turn, pray, pray, be a geologist!"[12]

As William grew up, his inquisitiveness blossomed into aspiration. He devoured every science book he could lay his hands on. But he also devoured the Bible, even engaging in lively recitation duels with his extremely religious father. The Church of England was a dominant force in his boyhood and would remain one his entire life. Buckland won a scholarship in 1801 to study for the ministry at the prestigious Corpus Christi College at Oxford University, where he never failed to miss a lecture on mineralogy and chemistry. During vacations, he fed his appetite for geology by carrying out field research on strata, traveling around England by horseback, even after becoming a minister.

Elected a fellow of the college—which entitles one to certain privileges such as dining at the best table for free—Buckland would go on to become the spectacularly eccentric first professor of geology at Oxford in 1818, when a readership was created specifically for him. Despite his insatiable appetite for knowledge about rocks, he spent much of his academic career trying to prove that religion and science weren't opposed to one another. Theology held more sway than science at Oxford, and virtually all of the important staff members were ministers of the Anglican church. Buckland's geological excursions and his passion for the controversial new field of study precipitated more than a few ripples in Oxford's serene waters. When he went on a tour of the Alps in Italy, one of the Oxford dons, Dean Gaisford, was heard to state "Well, Buckland has gone to Italy; so, thank God, we shall hear no more of this geology."[13]

It was around 1815, when Mary was just 16 years old, that Buckland, who was 15 years older, most likely first arrived at the Anning home, perhaps in top hat and academic robes. Almost nothing is recorded about their first encounter but, weeks before, Molly might have received an introductory letter in the post regarding Buckland's impending trip to Lyme Regis and a request that the young Miss Anning accompany him to the beach. Molly might very well have known of Buckland's family through his visits to the area as a young boy. But his arrival as a grown man, now a respected member of the Oxford circle, so much older than Mary, was likely to have caused a stir in a town where appearances meant everything. In general, a lower-class figure such as Mary normally wouldn't have mingled with a man in such a high position.

Although she probably was nervous, Mary's spirits surely would have soared at the news of Buckland's visit. No doubt she would have had to overcome the reservations of her worried mother, who would have been wary of allowing her daughter to leave the house with such a man unaccompanied. But in the end, the opportunity to go fossil hunting with someone from Oxford, the country's most prestigious institution, would have been too good for mother or daughter to pass up.

Descriptions of Buckland indicate that he was the kind of man people were instinctively drawn to. So commanding was his voice that students hung on his every word. Bounding into the room, a ball of intellectual energy, he was a bit like a fast-moving stage-coach: People felt they had to either jump on board or run out of the way. Graced with an agile mind, he was a great debater and a born experimenter who couldn't have cared less about what others thought of him. He had a round, jolly face and an outrageous sense of humor. His rooms at Oxford were notorious for being littered with cages housing every kind of animal imaginable. There were yellow snakes, green frogs, and chocolate-colored guinea pigs that were allowed to scurry freely across the floor like puppies. There also was a plethora of birds, whose high-pitched calls pealed out from the windows.

Buckland was so eccentric that, as a young man, he announced his intentions to eat his way through the animal kingdom.[14] He was known to serve his guests nausea-inducing combinations, such as mice on toast for supper, or even hedgehog, "good and tender." Once he amused friends by telling them that the only thing that tasted worse than a mole was a bluebottle fly.

His quirkiness carried over into the classroom, and in time his lively lectures attracted not only students but also senior members of the university. In later years, "Buckland liked, when lecturing on fossil footprints, to demonstrate how he imagined dinosaurs would have walked, an exhibition more resembling a flustered hen beside a muddy pond than the Reverend Professor of Geology at Oxford," novelist and Lyme Regis historian John Fowles wrote.

Mary surely had never encountered anyone quite like Buckland. He was something to watch, and, for Mary, he was someone to learn from. Unlike Buckland, Mary was pragmatic and realistic. She didn't have the luxury of being able to dally in studies of hedgehogs and horseflies but was forced to focus on finding interesting curios so that she could help her family survive. Any kind of meat was a rare treat. Her only pet during her lifetime was Tray, a black-and-white terrier-type dog of mixed breed, a stray Mary would pick up years later, as a young woman.

But, like Mary, Buckland had a trained eye. When he saw a tide coming in, he knew to grab his fossils fast. He knew that whatever wasn't retrieved quickly could slip away, disappearing into the sea forever. During their days on the beach, Buckland and Mary likely found all sorts of curios, ranging from fragments of jaws to entire rib cages. Whenever they noted anything interesting, a frenzy of picking and digging and chiseling ensued. Sometimes they were fortunate enough to come across an individual bone protruding from a cliff side. And sometimes they found nothing. More often than not, it was probably Mary who made the best discoveries. Although Buckland was an educated man, he hunted fossils only a few weeks of the year. Mary was out searching almost every day and knew exactly where to look and what to look for.

Many years later, Buckland's daughter wrote in her biography of him: "The vacations of his earlier Oxford time were spent near Lyme Regis. For years afterwards local gossip preserved traditions of his adventures with that geological celebrity, Mary Anning, in whose company he was to be seen wading up to his knees in search of fossils in the Blue Lias."

Buckland's trips from Oxford to Lyme Regis were always arduous ones, a 100-mile trek that would have taken many days for his favorite old black mare, weighed down with heavy sacks and hammers.[15] But the mare reputedly was a smart one and soon learned her duty, even seeming to take an interest in her master's pursuits. Friends said the horse always remained in place, without anyone having to hold her, whenever Buckland paused during a journey to examine a section of strata. She then stood patiently as he loaded her up with interesting but weighty specimens. Ultimately she became so accustomed to the routine that she eventually came to a full stop whenever even a stranger was riding her and they came across a stone quarry; nothing could coax her to move on until the rider had gotten off and at least went through the motions of examining the nearby rocks.

During her many days with Buckland, spread out over several years, Mary would have had the chance to pick his brain. Buckland was easy to talk to, and she would have learned a lot. Mary would have been sharp enough to catalog the minutiae of their talks as well as his observations on the curiosities they encountered. Although they were different in every way imaginable, their ardent enthusiasm for geology and paleontology was perfectly matched.

Following a day of collecting together, Mary would have made her way up the beach and back to town, through a gap between the buildings and onto Church Street, which was a short walk to her family's modestly furnished cottage. Most likely awaiting her was a bowl of gruel and some bread, without any butter. Buckland, however, would have bounded straight for the plush Three Cups Inn, where he would have warmed himself before a roaring fire while waiting for a glass of wine to be poured and a slab of marinated beef to be cooked. Afterward he would have climbed into an iron canopy bed with crisp white sheets and a firm mattress.

At times, Mary must have cast an envious eye toward Buckland. Even so, she appreciated their time together enormously. Mary's life had been a tortured one, but it was finally intersecting with the lives of others who had much to offer. Days with Buckland, as well as with De la Beche and other learned gentlemen, were presenting her with a richer world than she had ever dreamed possible.

4

A Great Kindness

On a crisp clear day just before Christmas in 1814, a controversial but sublime painting titled "Christ Rejected" went on display at the Royal Academy, a museum located on the Pall Mall in London, where, only seven years earlier, the first public street lighting with gas had been installed.

The painting was by Benjamin West, the first American artist to become internationally renowned. It depicts Christ being brought from the judgment hall by Pontius Pilate to Caiaphas, the high priest. With a white glow around his entire body and a shimmering halo surrounding his head, Christ exudes goodness even as everyone else depicted obviously wants him dead. Christ shows no sign of fear; indeed, his look of pity is not for himself but for everyone else. This monumental religious canvas was widely discussed, drawing a number of commentaries at the time—not all of them flattering—including a lengthy one published in an anonymous pamphlet which circulated in England in 1815. Some viewed the painting as absolving Pontius Pilate of guilt; others simply didn't like the title.

Mary must have seen the painting—likely a reproduction in a magazine, with commentary—because she transcribed pages of those comments into a small notebook.[1] By this time she was in the habit of keeping notes about what she read. In particular, she highlighted a passage on how the mob in the painting embodies not the failings of a single person but rather the thoughtless and "savage nature" of men. Mary also copied material on "the veneration and love for the accused" shown by the women of Galilee.[2]

Mary may have been enthralled for months, or maybe only a few days, by West's depiction of Christ, a painting that eventually found a permanent home at the Pennsylvania Academy of Fine Arts in Philadelphia. One can only imagine how the passages may have affected her. But clearly she was a contemplative girl who never enjoyed the luxury of being a frivolous young woman.

Certainly Mary never looked typical. Dressing for comfort rather than style, she almost always wore several layers of clothing, no doubt to ward off the cold that was often even prevalent in the summer months, as well as sturdy black boots, or clogs, and a battered high hat, possibly a precursor to the hard hat. The ensemble made her look heavier and even more masculine than she actually was. Most likely her hat was crafted of felted wool that had been repeatedly coated with shellac until extremely stiff—making it into the perfect protection against falling rocks. She also almost never left the house without a hammer in her right hand and a basket or a large collecting bag slung across her back.

While fashion-conscious girls were fussing over frilly ruffles, puffed sleeves, and the length of their gowns, Mary appeared not to care a whit about the day's popular cleavage-smashing corsets, most likely because they were expensive and so constrictive as to make bending over on the beach uncomfortable if not downright painful. She likely owned a hairbrush but most certainly couldn't afford the waxes, lotions, or rag rollers necessary to create the lovely little ringlets so popular with girls of the early nineteenth century. But Mary was not unattractive. She was blessed with a strong nose and chin, huge penetrating eyes, and a long dark mane of hair, generally tied up in a bun, hidden from view under her hat. She was a young woman who might have been pretty if, like other girls, she'd had the time or inclination to pay much heed to her looks.

Shortly after Mary found the skeleton of the ichthyosaur in 1813, a Mrs. Stock—who often had hired Mary as a girl to run errands for

her and had grown very fond of her blunt-tongued, unkempt proté-
gé—gave the 14-year-old the gift of a geology book, the first Mary
ever owned.[3] The grandmotherly neighbor, apparently with no children of her own, always saw to it that the Annings had something
to eat. She liked to say that Mary was a "being of the imagination."

By the time she was a teenager, Mary was a voracious reader,
often reading the same material over and over due to the unavail-
ability of books. Mary always seemed to be in pursuit of information, begging anyone who'd listen for help finding out what
the "bigwigs" in London were discovering about the new science
of geology and its even newer cousin, paleontology. If she managed to get her hands on a scholarly article by a geologist, she
was likely to spend many hours neatly transcribing it into one of
her many journals, painstakingly reproducing the illustrations as
well. Once she copied eight full pages of highly detailed illustrations from a paper on marine reptiles; those who saw it said it was
nearly impossible to distinguish Mary's copy from the original. By
all accounts, Mary was a relentless learner. Most likely she read
her first geology book so frequently that it became totally worn
out. Within a few years, she became a self-taught expert in the
arcana of anatomy, animal morphology, and science illustration.
Her quest for knowledge was insatiable. She soon exuded a confidence built on years of hands-on experience. Most important,
her finding of the ichthyosaur, as well as her finding of a number
of smaller fossils, was helping scientists come up with all the right
questions at a time when there were no obvious answers.

Her unerring eye for a fossil's best hiding places, developed
through hours of on-the-spot training, most likely made her
insights more valuable than those of the armchair theorists of
the time. Day after day, no matter what the weather, Mary was
a habitual presence on the shoreline, padding about in her long
skirts and shawls, toiling away at "man's work" against tempera-
mental skies and raucous tides.[4]

So proficient was Mary at fossil hunting that she was able to
scout out exactly where new fossils might be found after a storm.[5]
According to local lore, she developed such a nose for ferreting

them out that she would have been able to face 50 similar nodules and pick without hesitation the one that, when split with a dexterous whack, would reveal a perfect fish embedded in what was once soft clay.[6] After meeting Mary, science writer John Murray later recalled: "I once gladly availed myself of a geological excursion and was not a little surprised at her geological tact and acumen. A single glance at the edge of a fossil peeping from the Blue Lias revealed to her the nature of the fossil and its name and character were instantly announced."[7] Not only was Mary able to find fossils, but she also was able to remove them from the surrounding rock with the nimbleness of a concert pianist. The modern method for removing and gathering together a fossil from stone is to first use a plastic varnish to harden any exposed bones. But in those days, plaster and plastic were mostly unavailable; Mary wouldn't have had any real way of strengthening the bones, although she sometimes might have used a mixture of sand and hot wax. In addition, she simply mustered a world of patience and chipped away at the slabs of rock, one tiny sliver at a time, as cautiously as she could.[8] She would have placed the slabs within a wooden frame that could be mounted on a wall. Unless the specimen was small, most likely she transported her curiosities back home from the beach in several trips, one piece at a time—a part of collecting so laborious it would have put a serious physical strain on even the hardiest of lads.

Mary's penchant for discovery bubbled to the surface in the most bizarre and unladylike ways. When she wasn't trolling the beach, she was likely to be found studying not only long-gone animals but also modern ones, dissecting dead squid, cuttlefish, and other soft-bodied cephalopods right on the kitchen table. The more she examined them, the better sense she was likely to have of what they ate, how they lived, and in what ways they moved their bones and muscles.

Cephalopods are not the prettiest of creatures even when alive; they can change color, texture, and body shape as well as squirt out an ink in self-defense. None of this seemed to make Mary squeamish. When it came to squids, which have three hearts and bluish blood, Mary's first action would have been to flip them over so that she could find the noodles that look a bit like rigatoni in between their eyes. These are used for jet propulsion. With the siphons facing up, she would have pinched the mantles, which house the internal organs, and then cut along them carefully to the tip. Every female has two white glands, nidamental glands, smack in the middle, whereas every male has a white fluid-filled sac in the posterior end of the mantle. Dissection was hardly a fitting task for any person with a weak stomach.

Dissecting animals was a peculiar way for a young woman to spend her free time, but this was hardly the only unconventional behavior Mary was engaged in. In 1815, when Mary was 16, a teak sailing ship called *Alexander* was on its way from Bombay, India, to London when, at 2 A.M. on Easter Monday, it was swept up in a ferocious gale. The ship sank quickly not far from Lyme Regis. By morning, the body of a beautiful woman had washed ashore. Mary was accustomed to scouring the coastline for traces of the long dead, but the newly waterlogged body in fancy clothes would have jolted even her. The death of this woman, later identified as Lady Jackson, seemed to have touched Mary. The teen spent hours tenderly pulling seaweed from the lady's long hair and then, much to the bemusement of her neighbors, visited the woman's body daily at St. Michael's Church, where she would strew it with fresh flowers until it was finally claimed. The strange incident was recorded by Mary's close friend, Anna Maria Pinney. She wrote that Mary's daily visits to the corpse at the church exposed the "wild romance of her character." Most likely it was evidence of the fascination young people had with death at the time. In the early nineteenth century, accounts of triumphant endings—obituaries of young people who died tragically and heroically—were published in newspapers and magazines and were enjoyed particularly by teenage readers.

When she wasn't indulging her scientific passions or helping around the house, Mary most likely would have been at church.

As she grew up, she continued to be a firm believer in God and in prayer, and she likely saw his handiwork in the beauty of the shoreline. By most accounts, Mary wasn't one to skip services at the Independent Church, an association that would have continued to set her apart from peers such as the Philpot sisters, who were more likely to attend the Church of England. Although Dissenters couldn't hold public office or attend university, the Independent Church was a natural fit for Mary in some ways. Although many clerics feared geology, the Reverend John Gleed, Mary's pastor until 1828, supplemented his own meager salary by selling fossils and even competed with Mary for customers. With only two children left, Molly Anning was likely to have spent many an afternoon at the church, alongside Gleed; praying that no harm would come to her daughter. Like any mother, Molly would have been frightened for every minute of Mary's time on the rugged shoreline.

As a collector himself, Gleed likely was aware that fossil hunting was dangerous work. Landslides that happened so fast they could kill without warning were a constant threat. More than one person had lost his life on the famously wobbly cliffs; none stood a chance against a friable bluff once it started to crumble. In the past, strong landslips had sometimes caused such noise that townspeople mistook them for earthquakes. Boulders would break free and cascade down the cliffs and then out toward the sea, transforming the beach into a jagged maze of giant obstacles.

Of course, by now most collectors knew that the best time to look for fossils was after a landslip. Mary, however, seems to have had her own method for staying ahead of the competition—and it was a reckless one. Apparently she dashed out onto the shore when the waves were out, snatching up anything that looked even remotely worth keeping minutes before the next big wave had a chance to sweep everything away—herself included.

Unfazed by the hazards, Mary threw herself into her work. According to one famous collector, Thomas Hawkins, Mary explored the cliffs "when the furious spring-tide conspired with the howling tempest to overthrow her . . . and rescued fossils from the gaping ocean, sometimes at the peril of her life."

It is likely that living on parish relief for five years had caused Mary to focus her efforts even when she might have preferred to do otherwise. Often the family faced the real threat of starvation, and fossil hunting was the concrete means by which they knew how to earn a living. In addition, Mary likely continued her pursuits in an effort to honor her father's memory. His dream had been to open a proper fossil shop, one with a glass-fronted window through which he could show off his wares.

Mary's familiarity with the cliffs—honed by the many hours she put in on the beach—set her apart not only from those in Lyme Regis but also from the pioneer geologists from London and Oxford, who often visited Lyme to tap into her talents. As she matured from girl to woman, people in town must surely have thought her aura of celebrity was merely temporary. Surely she'd eventually be forced back into reality, compelled to take a job as a domestic or as a factory worker unless she found a husband first. But Mary was smart enough to realize that, without her fossil hunting, she was never likely to be much more than the destitute daughter of a deceased Dissenter carpenter—just another girl of low rank, never formally educated and never likely to go anywhere. On the beach, or in her father's workshop, she would have felt emancipated from the confines of her station in life. By nurturing her talents, she was able to attract the kind of company that was unheard of for other girls in her situation.

Between about 1815 and 1819, when Mary was not even yet 20, she unearthed several more complete specimens of ichthyosaurs, allowing geologists like Buckland to glean a clearer image of the creature's anatomy. Some she uncovered were only the size of a trout; others, nearly as big as a baleen whale. No matter what the size, though, what likely always struck her the most were the enormous round eye sockets. They made the beasts look devious, even evil. For scientists, Mary's skeletons provided a picture of the kind of creatures that inhabited the seas during the remote past. In Mary's time, the seas were teeming with fish; the oceans of yore, however, must have been overflowing with reptiles. For geologists, the ichthyosaur specimens raised the possibility that a link once existed between fish and reptile.

Mary's discoveries brought her in touch with a huge cast of geological characters, some quite eccentric. One was William Daniel Conybeare who, like his good friend Buckland, also became interested in geology at Oxford and also took holy orders. His father had died, leaving him with plenty of savings. Another geologist, the Geneva-born Jean Andre De Luc—the scientist who supported the Genesis account of creation—first began exploring the geology of the Dorset coast in 1805. He was a frequent visitor at Lyme Regis until his death in 1817.

By 1816, Mary's close friend Henry De la Beche had set off from Lyme Regis on a tour of the north of England and Scotland with Thomas Carpenter, the Lyme doctor and coroner, and George Holland, a meteorologist.[9] The trio visited various sites of geological interest, in particular a part of Aberdeenshire where they studied flesh-colored granite, then the subject of heated discussion among geologists. Some explained its formation by volcanic action while others, such as Abraham Werner, continued to believe that all rocks were shaped by a global ocean. By all accounts, De la Beche honed not only his geological skills but also his artistic talents during this journey, completing several watercolor sketches of landscapes and fossil finds.

De la Beche was able to focus exclusively on his personal interests thanks to his healthy financial situation. In 1817, when he turned 21, he began receiving a hefty income from his family's Jamaican estate. Even though this income was always at risk, especially as the campaign to abolish slavery grew in momentum, it would be a long while before De la Beche would be forced to actually work for a living.

With his confidence growing, De la Beche was even able, upon returning from his travels, to join the prestigious Geological Society of London.[10] His first major contribution to the group's proceedings was a geological memoir on the southern coast of

England. Also upon his return, and perhaps at his mother's prodding, Henry proposed to Letitia Whyte, the pretty daughter of the highly regarded Captain Charles Whyte. News of the impending marriage probably would have caught Mary completely off guard. With most people believing that marriage was the only way out of poverty for a woman like Mary, her friendship with De la Beche had been the source of at least some speculation. And now, at least with this particular close friend, she'd lost her chance.

Two years later, in the summer of 1819, when Mary was just 20, De la Beche whisked his new bride and her mother, along with a contingent of servants, off on a tour of the Continent, which included a brief residence in Switzerland. Even on his honeymoon, he made time to befriend the leading naturalists in every country they visited. Upon their return to Lyme Regis in 1820, De la Beche published his findings on alpine geology and within a year issued a joint paper with Conybeare called "Memoir on the genus *Ichthyosaurus*," which described in great detail many of the fossil remains that Mary had discovered over the years.

With De la Beche busying himself with a new bride, Mary began spending more and more time with the Philpot sisters.[11] Mary grew particularly fond of the fiercely independent Elizabeth, whose regard for geology was just as strong as her own. Soon, working together, the Philpots had amassed quite a formidable collection of curiosities. Mary was only too happy to help add to her friend's assortment by seeking out ammonites, which were easy to find near Monmouth Beach to the west of Lyme Regis, as well as belemnites and arrowheads, pyrites, and cockles. A local woman, Selina Hallett, remembered the Philpot collection this way: "Several cases with glass tops and shallow drawers all down the front stood in the dining room and the back parlor and upstairs on the landing, all full of fossils with a little ticket on each of them."[12] The Philpot "museum" soon became a perennial favorite with the prominent geologists passing through town.

Not only were the Philpots known for their fossils, but they were also famous, at least locally, for their homemade soothing salve.[13] Hallett wrote that the sisters were always kind to poor people, supplying their salve to anyone who wanted it. And there were a lot of people who would have wanted it. No one was exactly sure what was in the salve, but it most likely was some combination of beeswax and sweet oil melted together with a dash of turpentine; or perhaps it was a combination of lard, resin, tobacco, and beeswax. Whatever it was, the salve was doled out for just about everything: burns, frostbite, abscesses, chapped hands, bug bites, and fever sores.

Most important for Mary, though, was that, in Elizabeth, she had found a friend with whom she could share her geological passions. At least a few times each week, the pair strode along the shore, bouncing ideas off one another whenever they came across something unique. Although nearly 20 years older than Mary, Elizabeth often asked for her opinion and she forever encouraged the younger woman's pursuits. Their conversations probably were a welcome tonic during times when Buckland, De la Beche, and other learned gentlemen were off in London or Oxford or gallivanting around the country or the Continent, searching for new treasures.

Like Mary, the affable Buckland also was throwing himself into the study of geology between 1815 and 1819, but mostly from within the traditional confines of Oxford University. Unlike Mary, he had a self-imposed purpose behind his passion. Above all else, Buckland was determined to convince the world that geology had the ability to enhance religion, not undermine it. By 1819, interest in his lectures had grown so great that a readership in geology was created at Oxford, with Buckland being the first holder of this prestigious appointment. As such, he was required to give an induction speech, set for May 15. Unbeknownst to his audience, though, he planned to use this venue to take his first real stab at reconciling geology with the Scriptures.[14] For all his aplomb as a speaker, Buckland's nerves apparently were always rattled, initially, by the sight of a new audience. His jitters calmed only after he made someone in the audience laugh. But there wasn't likely to be a lot of laughter during this particular address. Lecturing in the esteemed halls of Oxford University on such an occasion would have been a stiflingly formal affair.

The young students, so dignified in their swirling long gowns, would have filed into the room first, avoiding the front rows, which were set aside for academic staff. As Buckland peered around him, he might have been flushed with apprehension, especially when noticing that some of the university's highest-ranking officials were in attendance. But once introduced, he organized his notes, stood up confidently, and launched into his discourse.

One by one, he laid out the facts as he saw them, arguing that Earth was created by God an incredibly long time ago, perhaps hundreds of thousands of years in the past. The considerable thickness of primary rocks—or the rocks that apparently were formed first and were free of fossils—was proof that in the beginning the world was devoid of life. As spelled out in Genesis, God created different creatures at different times, saving the best for last: humans made in his own image. Corroborating the Bible's text was the succession of different kinds of fossils in the geological record. Primitive creatures, such as reptiles, were always found in the lower strata, while those less primitive, such as mammals, were always found higher up. Buckland noted how the country's most valuable mineral resources, like coal, were the most prodigious ones. God also had helpfully located them in places where they could be extracted fairly easily. If anyone needed further proof of the creator, they only had to look at the miraculous design of all living organisms, each one forming a link in the boundless chain of all earthly beings.

Quoting from *Natural Theology,* a popular book written by Christian philosopher William Paley, Buckland said that "every example of natural design, from the movements of the planets to the interlocking mechanism of the barbs of a bird's feather, were incontrovertible proofs of the existence of the supreme intelligent author."[15] Concluding his lecture, he introduced his hypothesis that the word *beginning* as used in Genesis expressed an undefined period of time between the origin of Earth and the creation of its current residents. During this period there was a series of upheavals, with successive creations appearing and disappearing, culminating in the arrival of modern-day animals and humans.

When Buckland finished speaking, there might have been a few moments of unsettling silence. But surely it was followed by spirited applause. Undoubtedly he would have stepped down with a satisfied smile on his face, strode back to his seat, and enjoyed congratulatory handshakes with those around him. By most accounts, the response was rapturous from an audience most likely relieved that the geologist hadn't tried to take swipes at biblical teachings. Instead, Buckland had praised God's benevolent intentions. He had emphasized the validity of geology—but mostly as a means of validating religious truths.

Most personally satisfying for Buckland was the fact that he'd successfully justified, at least for the time being, the importance of the new science of geology, meaning he was free to go on with fossil hunting with the church's blessing. This was vital as Buckland had become increasingly obsessed with the incompleteness of the fossil record, a deficiency Charles Darwin lamented some 40 years later. With no skeletons of dinosaurs or other creatures, the fossil record largely was drawn from a series of ichthyosaur finds—many of them provided by Mary.

Since Mary Anning's first great find of the 17-foot skeleton, geologists had been more than happy to accept any and all of her specimens, but they were not always willing to give her credit. When they published their findings, her name wasn't included. Even when her first ichthyosaur was cited in scholarly journals, her part in its retrieval was omitted. She also obligingly escorted a steady stream of visitors along the shoreline—but with little recompense.

It was 1820. De la Beche was married, and a 21-year-old Mary was still living with her mother and brother in the same small cottage in the center of town, overlooking the sea, always with a table of curiosities set up outside. Many scientists were familiar with her name and reputation. But her family's finances remained in dire straits, as did those of most people around them.

The war with France was long over but Napoleon's defeat at Waterloo in 1815 had sunk Dorset's agrarian economy into a deeper depression. When it no longer became necessary to fund a large army and navy, the government abolished the income tax that had been levied on the rich and instead increased the indirect taxation on food and other necessities that was paid primarily by the lower classes. Before the end of the decade, half a million people were unemployed, fueling tensions among the working class, middle class, and upper class that often led to violence. In December 1816, thousands of protesters had gathered in Spa Fields, just north of London. After breaking into a gunsmith's shop, they marched on Westminster—the seat of British government—in a show against high prices. The army stopped them, arresting 300. Three years later, tensions flared again. This time, in a famous incident known as the Peterloo Massacre, cavalry charged into a crowd of 70,000 protesters gathered at St. Peter's Field in Manchester, killing 15 people and injuring hundreds of others, including women and children.

Lyme Regis wasn't prone to such protests. But poverty was luring some residents into another unsavory activity, one that went on at night, while Mary and the rest of the town slept. That activity was smuggling.

Also called "free trading," smuggling may have been distasteful, but it also was helping to shore up the local economy in the early 1800s.[16] At Lyme Regis, cargo unloaded at the Cobb could easily be diverted up side alleys to bypass the customs house. Contraband, mostly brought in from the Channel Island of Alderney but sometimes from the northern coast of France, included casks of tea, silk, brandy, and tobacco. Any time smugglers were being hunted by customs officials aboard revenue cutters, they could simply rope the casks together in a raft and sink them offshore. They would mark the casks' position by a float so that they could retrieve the casks later by "creeping"—fishing them up using grappling hooks. If smugglers didn't have time to sink a raft, they could throw the kegs overboard. To secure a conviction, both smuggler and contraband had to be caught. By separating themselves from the contraband, smugglers increased their chances of escape.

Mary was greatly sympathetic toward the smugglers. When she ran across a keg of contraband along the shore, she attempted to cover it up to hide it from the excise men, who were responsible for collecting taxes. Often she would tell some deserving family where to find the hidden cache. Because of the taxation on imported goods, people in Lyme Regis considered smuggling almost a "legitimate" trade. For many, it was the only way they could provide food for their families.

Fortunately for Mary, in these wretched times, when many of Lyme Regis's other residents were becoming endlessly creative when it came to making ends meet, she somehow managed to draw the attention of a very important admirer. It was 1818 when Lieutenant Colonel Thomas James Birch first entered Mary's world.[17] This well-to-do fossil collector, a retired officer in the Life Guards, was touring England's West Country, spending his half-pay pension in the search of fossils. Often he went to Lyme Regis, where he took to visiting with Mary and her mother in their home, buying many fine specimens from them.

Shortly after their first encounter, Mary discovered a nearly complete ichthyosaur, and Birch purchased it. Mary would have been awed by Birch's cultivated, articulate speech as well as a bit amused by his reserved demeanor. In the rough-and-ready world of the Lyme Regis harbor, Birch was probably something of an elegant white knight.

One winter day, about a year later, Birch showed up at the Anning home, only to find the family in a state of real distress. For the past several months, Mary had not had one lucrative fossil sale. Parish support, too, had finally come to an end. By this time, Molly might have been worryingly thin, edgy, and willing to talk frankly about the family's predicament. In short, they were so desperate that they were making plans to sell off their furniture just so they could make their rent.

Most likely Birch listened to Molly's woes and was livid. The Annings had made great contributions to the geological community. How could they be destitute? Unable to ignore their plight, he decided he must do whatever he could to get them some funds—and fast. Birch explained their situation—as well as his outrageously generous solution to get them out of it—to a fellow geologist, Gideon Mantell, in a letter dated March 1820:

> I have not forgotten my promise to select for you some fine things from the Blue Lias—I cannot however perform it yet as I have great occasion for every individual specimen I can muster. The fact is I am going to sell my collection for the benefit of the poor woman Molly and her son Joseph and daughter Mary at Lyme who have in truth found almost all the fine things, which have been submitted to scientific investigation.... I may never again possess what I am about to part with; yet in doing it I shall have the satisfaction of knowing that the money will be well applied. The sale is to be at Bullock's [London museum] in Piccadilly the middle of April.[18]

And what a sale it was. The title of the catalog—"A small but very fine collection of organized fossils, from the Blue Lias Formation. At Lyme and Charmouth, in Dorsetshire, consisting principally of Bones, Illustrating the Osteology of the Icthio-Saurus, or Proteo-Saurus, and of specimens of the Zoophyte, called Pentacrinite, the genuine property of Colonel Birch, collected at considerable expense, which will be sold by auction by Mr. Bullock, at his Egyptian Hall in Piccadilly, on Monday, the 15th day of May, 1820"—failed to do it justice. In total, Birch's trove included 102 items for the auction block. Highlights of the sale included: "Fragment of a fish, the reverse side of the scales beautifully seen," "part of the foot of a prodigiously large animal," and concluded with "a most interesting illustration of the osteology of the Icthio-Saurus, or Proteo-Saurus." For the most part, the items weren't just hastily chipped-out trilobites but were exceptional in their peculiarity. Many of the more unusual items were those Mary had prospected herself.

The Egyptian Hall in Piccadilly was the perfect place to stage an auction of this magnitude if the goal was to build hype. Since the turn of the nineteenth century, a mania for all things Egyptian had swept through fashionable Europe. And London was no exception. Decked out with Egyptian friezes, the hall had become the city's most illustrious place to hold an exhibit as it was exotic in its decor, both inside and out.

It was no wonder Birch's fossil auction drew record interest from all across Europe. The three-day sale started precisely at 1 P.M. The procession of his beloved items moved quickly, creating something of a party atmosphere, with specimens snatched up by the highest bidders from Germany, Austria, all around England, and even by one from France—Georges Cuvier. Some of the finest lots went to the Museum of the Royal College of Surgeons in London, such as lot 102, the Anning/Birch Ichthyosaur, which fetched a gratifying £100.

In total, Birch's collection brought in more than £400—comparable to nearly $50,000 today—all of which he handed over to the Annings.[19] The event also delivered a much-needed publicity boost, and on an international stage, to the Annings' beleaguered fossil business. Seemingly unemotional about the sale, the gallant Birch merely appeared happy he could be of assistance. His fossil collection had been a wonder of the geology world, and the gesture gave Mary some star status. Mary, Molly, and Joseph were incredulous. Never before had anyone shown them such kindness and, for the first time in their lives, Mary and her family were financially secure.

Across Europe, an increasing number of fossil collectors began asking about this young woman from Lyme Regis named Mary Anning, the young recipient of the auction's proceeds.

Birch's act of generosity sparked some fantastic rumors. Shortly after the auction, one major fossil collector, George Cumberland, noted that "Col. Birch is generally at Charmouth…they say Miss Anning attends him." He was referring to the town just a few miles from Lyme Regis, between which stretched a sandy beach, perfect for fossil hunting. At the time, Birch was 52, over twice Mary's age.

5

A Long-Necked Beauty

Word of Birch's generosity tore through Lyme Regis and, forever after, no one looked at Mary in quite the same way again. Until then, many people had jealously dismissed her as a bootlegged version of a real fossil hunter—someone who simply got lucky by stumbling across some old bones. Charles Dickens later wrote that, in her own neighborhood, people "turned and laughed at her as an uneducated assuming person, who had made one good chance hit."[1] After the sale, though, at least some of these townspeople would have gazed at her with a new kind of respect, smiling deferentially each time they crossed paths. But there was likely to be small-town whispering too. No doubt people were wondering what a dignified bachelor in as high a position as Birch possibly could have been thinking when he sold all the fossils he owned, and such a fine collection at that, simply to benefit a family with whom he had no real relations. The young Mary was clever, for sure, but she was uncouth and not the least bit feminine; what could he possibly see in her?

If Mary minded the gossip, she wasn't the type to show it. More than likely she was basking in the unexpected attention and publicity. Mostly, though, she just got on with business, but not before allowing herself a few indulgences first. There are some indications that she bought herself a pretty high bonnet with the money from

Birch—one made for fashion, not function.² Bonnets were popular with women well into the mid-1800s. Mary's bonnet most likely mimicked the top hat and was made from cloth decorated with ribbons, feathers, and flowers. Meanwhile, Mary's neighbor George Roberts was writing the first complete history of Lyme Regis, and he needed subscribers in order to pay for the printing. Using the money from the Birch's sale, Mary placed one of the initial orders.

Despite the advantages of suddenly receiving a small fortune—enough to finally put some real food on the table as well as to set aside savings for future use—Mary was never one to remain idle. A year after the auction, in early 1821, perhaps by this time accompanied by Tray—always the reliable guard dog—she discovered and excavated a beautifully preserved ichthyosaur only five feet long. Mary immediately wrote of the find to Henry De la Beche, who offered it to the British Museum. That same month she spent days digging out another much larger ichthyosaur skeleton, this one a fearsome 20 feet long, which later was deemed to be the first flat-toothed *Ichthyosaurus platyodon*. Later that year she found another five-foot fossil that was eventually named *Ichthyosaurus vulgaris*.³ Early in 1822 she retrieved yet another large ichthyosaur, this one at least nine feet long. As before, these finds were credited not to her but to the monied gentlemen collectors who purchased them.

Scientists were beginning to grasp the subtle differences between the ichthyosaur specimens Mary was finding, which led them to conclude they didn't belong to the same exact species. De la Beche and his friend and fellow geologist, the Anglican clergyman William Conybeare—now also a member of the Geological Society of London—deduced from the teeth alone that there must have been at least four disparate species of these magnificent creatures, all superbly designed to be skilled hunters and killers. But despite her success at uncovering ichthyosaurs, Mary hadn't been able to track down anything truly new or newsworthy. While Birch's money may have gone a long way toward easing the family's worries, it was a financial cushion that wasn't going to last forever.

At one point Mary's mother, Molly, stepped in, taking the audacious step of begging Charles Konig, curator at the British

Museum, to pay what he owed for an ichthyosaur he had purchased. She wrote in a letter dated September 1821: "As I am a widow woman and my chief dependence for supporting my family being by the sale of fossils, I hope you will not be offended by my wishing to receive the money for the last fossil as I assure you, Sir, I stand much in need of it."[4]

Indeed, by 1823, when Mary was 24, the family's financial situation was once again precarious. The fossil business was an unpredictable one and, as geology grew into a popular science, there were more people hunting for fossils, meaning more competition.

Early in the year, Mary sold an exceptionally well-preserved ichthyosaur skeleton to a collection of geologists, who then presented it to the new Bristol Institution for the Advancement of Science—the first time the museum had ever received such a donation. During a lecture at the institution, the specimen was dramatically described as "the most valuable in the kingdom." But the lecturer failed to mention the young woman who had found it.

This time, however, the oversight did not go unnoticed; local geologist and fossil collector George Cumberland came to Mary's defense. He immediately fired off a letter to a Bristol newspaper full of praise for the "persevering industry of a young female fossilist of the name of Hanning." (He spelled her name the way local Dorset residents pronounced it.) Cumberland went on to share with readers how Mary had pinched from the cliffs "relics of a former world...at the continual risk of being crushed by the suspended fragments they leave behind....[T]o her exertions we owe nearly all the fine specimens of Ichthyosauri of the great collections."[5]

Mary was grateful for the acknowledgment. Indeed, she probably needed all the validation she could get after receiving so little recognition for her fossil discoveries, which had advanced scientific progress for so long.

By this time, as she approached her mid-twenties, Mary surely realized that her chances of marrying were growing slim. She must have wondered about Birch's intentions, but it was never clear if the much older fossil collector ever had any romantic interest in her. Perhaps, not having any family of his own, his motivations had been strictly paternal. Like her mother, Mary was all too aware that, in England's male-dominated society, there were few options for unmarried women of the lower classes such as herself. Her only alternative was to make her own way. And the only way to do that was to make another extraordinary find—the kind that would send tongues wagging across the scientific world.

So it was with a palpable sense of anticipation that this plucky young woman set out following a storm on the blustery tenth day of December in 1823. As always, she probably was quite a sight, by this time a driving perfectionist who never failed to engage with every single mound of clay, dirt, and rock she came across on the shore. Working at the foot of the menacing Black Ven, she first would have scanned the cliff face, trying to decide which patch of earth looked most promising. And then, quite suddenly, something would have caught her eye.

It was an object—sleek, glossy, and circular—poking out from an otherwise ordinary cluster of dark-gray mudstones. Most likely Mary rushed over and scraped away the surrounding shale as quickly as she could. There was plenty of it. After a while, though, a creature's skull emerged. But it wasn't elongated or snouted or pointy like the skull of an ichthyosaur. There weren't any huge bony cavities for eyes either. Probably she studied this new find in silence, with only the sounds of birds and waves and her trusted dog, Tray, to keep her company. Her heart surely raced. Just minutes before, she might have felt so chilled that she wondered whether it was worth staying outside. Now she was sure she'd come across something truly sensational, but there was an incoming tide on the horizon. She would have whirled around in a panic. She needed help—and fast.

By this time, everyone in town was familiar with Mary's work. It wasn't hard for her to enlist the aid of a few villagers in digging out the creature. Soon they were slogging away alongside her, into the evening and then—after a respite—returning the following morning. But rewards were forthcoming. Out came the vertebrae, pelvic bones, and—one by one—a series of ribs deeply embedded in the cliff side. First there were 4, then 6, and then finally 14. A bit more digging and the fine bones of what looked to be four legs—or might they be paddles?—also were unearthed.

The creature was starting to resemble not a crocodile but something akin to a turtle with a flat mouth and stubby short tail and, oddest of all, an abnormally long neck. Mary toiled away at the cliff, hour upon hour, chipping away with hammer and chisel, while waves lashed at her hefty petticoats and numbed her fingers. Once she started cutting away at the matrix imprisoning a particularly large fossil, it could take what seemed like forever to make any progress. But she was seeing some results, especially as she likely had others to help her. In the end, after several hours of jabbing away at the mud and rock and shale, a skeleton emerged that was about nine feet long and six feet wide, but with a head that was only about four to five inches in length. Surveying her find, Mary must have marveled at the peculiarity of the skeleton, which didn't boast legs or fins but indeed paddles consisting of many fine, dainty, delicate bones.[6]

This wasn't the first time Mary had come across remnants of this type of creature. But never before had she found a skull, and, what's more, the bits she'd found in the past had been so brittle they had very nearly withered away by the time she'd gotten them home. Now she had before her a much more substantial find—a complete skeleton. And it was a beauty.

By then, Mary would have known from De la Beche and Buckland that fossil experts had long suspected that in addition to ichthyosaurs, a second kind of sea monster had once commanded the ancient oceans. Conybeare, 12 years older than Mary, had become acquainted with De la Beche in the Assembly Rooms, or public halls, of Lyme Regis, and they had remained good friends. While helping De la Beche prepare his own written comments on the ichthyosaur in 1821, Conybeare had urged him to mention some bones that bore little resemblance to those of the ichthyosaur. Indeed, it was Conybeare who, while visiting Somerset, England, had come across a few flat-ended vertebrae that he deduced belonged to a creature nothing like an ichthyosaur. He also had found a jaw in which conical teeth appeared to have rested in sockets, in addition to a badly damaged skull. Raising even more questions were a mishmash of paddle bones. Although he had amassed only a jumble of fragments from various collections, Conybeare felt so confident that he was on to something at the time that he even put forth a name for the unknown beast: *Plesiosaurus,* meaning "near to reptile." Nevertheless, his enthusiasm was tempered by his own suspicions that he might have concocted a fictitious creature from the juxtaposition of bones belonging to a variety of species.

As Mary surveyed the strange creature before her, she wondered if this could be the entire skeleton of the creature Conybeare had speculated about. When word of Mary's discovery finally reached Conybeare, who was becoming increasingly well known as one of the founders of the Bristol Philosophical Institution in 1822, the rector was so excited that he was unable to finish drafting his Sunday sermon. He simply couldn't wait another minute before writing a euphoric letter to De la Beche, who was in Jamaica at the time: "Buckland...brought important news—that the Annings have discovered an entire Plesiosaurus." Three days later, Conybeare received what he called

a very fair drawing by Miss Anning of the most magnificent specimen....It was the evening also of our Philosophical Society at the Bristol Institution and you may imagine the fuss this occasioned. My sermon, though finished in scraps was then not half-transcribed, but one of my sisters-in-law, who was staying with me, kindly undertook the task, and to the Society I went....Such a communication could not fail to excite great interest; some of the folk ran off instantly to the printing office, whither I was obliged to follow to prevent some strange blunders...thus I did not get home till midnight.[7]

Conybeare was worried that in the commotion, people had rushed off so fast they were likely to spread misinformation. It was simply another fossil of a skeleton but, judging by the excitement it was generating, it might as well have been an alien from the moon.

News of the find spread fast and exponentially, all the way to France. But after perusing Mary's drawings, the eminent Georges Cuvier at the Muséum National d'Histoire Naturelle expressed suspicions that the new animal might be a sham. The length of the neck seemed impossibly unrealistic. Indeed, the volume of air needed to fill the trachea of such a drawn-out neck would have been extraordinary. To Cuvier, the animal's characteristics broke the almost universal anatomical law that restricts the number of cervical vertebrae, or neck bones, to no more than 7 in animals that walk on four legs. In birds, the number of cervical vertebrae is greater, typically varying between 13 and 25; living reptiles typically have from 3 to 8. Yet Mary's allegedly reptilian creature apparently boasted 35 vertebrae in the neck alone. Cuvier simply couldn't accept such a possibility.

To the scientific community, Cuvier was far too good at what he did to be wrong. Europe's star anatomist, Cuvier was famous for describing whole animals after examining only a single bone from their skeleton. With a long and distinguished history, he stood several rungs above other anatomists and certainly above other anatomists in England. Almost single-handedly, he had founded comparative anatomy—or vertebrate paleontology—as a scientific discipline. By this time he was also noted for his distinctive division of animals into four main branches: Vertebrata, Mollusca, Articulata (arthropods and segmented worms), and Radiata (cnidarians and echinoderms). He was the kind of authority who never had to prove anything; in this case, just his expressions of doubt were enough to cast a shadow over Mary's discovery. If he was dubious, then everyone else would be too.

Cuvier had held little esteem for the work of English anatomists since his dealings with Sir Everard Home, the "Surgeon to the King," whose inept misidentifications of Mary's first ichthyosaur had drawn scorn across Europe. It was easy for Cuvier to assume that the so-called English experts were being hoodwinked by just another fossil-collecting layperson. Perhaps the Annings had taken the head and neck of a sea snake and juxtaposed them onto the body of an ichthyosaur. As if to underscore his reservations, he highlighted the blatant location of a crack in the bone at the base of the neck as possible proof of a hoax. Cuvier wrote to Conybeare warning him that the creature could be a deception.

For Mary, Cuvier's misgivings surely were a disaster. If he convinced others that the new fossil was a forgery, the Anning family's reputation could be ruined forever. For years, amateurs had been known to take liberties with their finds in an effort to dupe professionals into exaggerating their worth. The English collector Thomas Hawkins, for instance, was a master of deception. Just before a 25-foot ichthyosaur from Hawkins was to be placed on display at the British Museum, curator Charles Konig had given it a careful examination.[8] It had looked perfect. In fact, it had looked too perfect. Konig checked the catalog and discovered that the very real-looking right front fin was merely outlined in the catalog, indicating it was missing. A large piece of the tail also wasn't to be found in the catalog's version. All in all, the "fossil" he was inspecting was largely plaster, although the museum had been under the impression it was genuine bone.

Gossip quickly arose about the beast from Lyme. Had it been fabricated? A special meeting to arbitrate on the matter was convened at the Geological Society of London, most likely in January 1824. Mary wasn't asked to attend. Most members were aware of and even sympathized with Cuvier's qualms; they were themselves stumped by how any animal could compensate for the weakness that would have resulted from such a lengthy neck. But as the gentlemen debated the creature's improbable combination of characteristics, they recognized that its features precisely matched all the fossil findings that Conybeare, with De la Beche's help, had made earlier in Somerset, findings that had led Conybeare himself to propose the existence of such a long-necked beast.

After lengthy and often heated debate, these earlier findings eventually convinced the society members that Mary's skeleton wasn't a fake. By the end of the evening, the Annings were vindicated, and, perhaps for the first time, Cuvier was shown to be fallible. Later, after more careful study of Mary's drawings and eventually the bones themselves, Cuvier openly admitted that he'd rushed to judgment and made a mistake.

While never one to doubt her own abilities, Mary likely breathed a huge sigh of relief upon hearing the verdict. Soon plans were being made for Conybeare to present a definitive scientific paper on the *Plesiosaurus* at a meeting of the Geological Society of London on February 24, 1824.[9]

This meeting would mark the first time Buckland presided as president—a role he took on with great pride that year. He couldn't wait. Adding to the distinctiveness of the occasion, he had arranged for Mary's discovery to be shipped to London, so that it could be showcased as part of a temporary exhibition at the society. At the same time, he had requested that Conybeare come to London a few days early to meet the skeleton; neither wanted it to fall into the hands of the ambitious Home, especially with his reputation for ineptness. Mary, of course, was not invited.

Buckland had been charged with the care of the plesiosaur by Richard Grenville, the first Duke of Buckingham.[10] The duke had agreed to pay Mary £110 for the skeleton—the highest price ever commanded by a single fossil—and he was eager for Buckland to study the specimen so that he could put his mind to rest by confirming the creature's identity. The duke readily agreed to ship the fossil to the Geological Society for Buckland's examination.

Buckland was hoping the plesiosaur would arrive in time for the society meeting, an awesome prop as Conybeare delivered his address on its anatomy. But the ship carrying the cargo was delayed in the English Channel. Even so, no one was unduly concerned; Conybeare's presentation, elucidated by Mary's professional and vivid illustrations of the skeleton, would bring the creature to life.

That night, the meeting room was heaving with an enthusiastic audience. Many members had invited friends after hearing rumors they were in store for a major announcement. The crowd sat quietly, in rapt anticipation, as Conybeare stood up to deliver his paper: "On the Discovery of an Almost Perfect Skeleton of the Plesiosaurus."

By most accounts, Conybeare never had been as entertaining a public speaker as Buckland, but his audience on that night was captivated just the same. Standing beside Mary's detailed drawings, he showed how the new discovery corroborated several of his conjectural points on plesiosaurian anatomy. The neck was as uncannily long as he had predicted. In fact, it was even longer. Initially, he had estimated there to be 12 neck vertebrae, but the new specimen had 35. All in all, it was likely to be the most monstrous creature that had yet been found amid the remnants of an ancient era—a description even Cuvier eventually agreed with. What helped him to draw this conclusion, Conybeare pointed out, was the fact that, after being locked away from the world for perhaps thousands of years, the fossil remains were as perfectly intact as the bones of animals currently in existence.

Conybeare went on to paint a picture of a creature that

> swam upon or near the surface, arching back its long neck like the swan, and occasionally darting it down at...fish...It may perhaps have lurked in shoal water along the coast, concealed among the seaweed, and raising its nostrils to a level with the surface from a considerable depth...a secure retreat from the assaults of dangerous enemies...the length and flexibility of its neck may have compensated for the want of strength in its jaws and its incapacity for swift motion through the water.

What Conybeare wasn't able to know for sure was that, over time, plesiosaurs would prove to be even more wondrous than ichthyosaurs. Indeed, despite their very unfishlike shape, plesiosaurs eventually replaced ichthyosaurs as the top aquatic predator in the Cretaceous Period, which ended about 65 million years ago. With their four oarlike appendages, the ferocious marine reptiles with razor-sharp teeth could quickly row through the water, much like penguins. At the time, Buckland memorably described the plesiosaur as a "serpent threaded through a turtle." Later, the Loch Ness monster would be described as a plesiosaur.

It was Mary's discovery of the plesiosaur that gave impetus to serious contemplations on evolution, which would years later feed into Darwin's theories on evolution. Although the ichthyosaur was the first extinct animal known to science, it wasn't completely unlike modern dolphins and tuna. But plesiosaurs were so different from any modern animal that they couldn't be so easily dismissed as a variety of a known existing creature.

With the audience still hanging on his every word, Conybeare concluded his address with the dramatic announcement of the creature's name: *Plesiosaurus giganteus*. Surely the audience would have been effusive in its praise, and likely he was delighted. He thanked the duke for allowing scientists to study the fossil and he praised the "scientific public" for the discovery—but never mentioned Mary's name.[11] The *Bristol Mercury* reported the find as well as the duke's purchase of it but also made no mention of Mary.[12] As time went on, whenever Conybeare was forced to refer to Mary, he generally used the word "proprietor," a somewhat negative description that inadvertently suggested that money was the only motive for her preoccupation with fossils.

In hushed tones, the esteemed members of the Geological Society most likely marveled among themselves at news of the creature's numerous neck vertebrae, imagining what it must have been like to see such a creature swimming close to the surface, its long neck rising high above the water. Yet members were in store for even more revelations.

Indeed Buckland almost immediately eclipsed the excitement over Conybeare's lecture, rising and charging that he had his own announcement to make. The eloquent man went on to describe his discovery of a creature he called *Megalosaurus*—or great lizard—in consultation with Conybeare.

The colorful Buckland began his address:

I am induced to lay before the Geological Society the representations of various portions of the skeleton of the fossil animal discovered at Stonesfield, in the hope that such persons as possess other parts of this extraordinary reptile may also transmit to the Society such further information as may lead to a more complete restoration of its osteology.[13]

As he went on, conjuring up an image of an increasingly disturbing ancient world, he told the stunned members how, beginning about 1815, he had begun acquiring the hoard of large fossil bones from the Stonesfield quarry near Oxford. Initially, he had no clue as to what animal the bones belonged. He turned the grab bag jumble of bones—a huge jawbone with teeth in place, several vertebrae and shoulder bone fragments, and some limb bones—over to the Oxford Museum, which was happy to have them. The more the bones were studied, the more they seemed to pertain to the Sauria, an order of reptiles including lizards.

He explained that, unfortunately, evidence was still lacking. Nonetheless, from the teeth alone, he had concluded that the ancient beast was not a mammal but a reptile. One reason for this conclusion was that the bones were found in the same vicinity of rock as marine animals as well as the remains of tortoises and crocodiles. Therefore, he reasoned that the creature was "probably an amphibious animal"—and one that could slink about both on land and in the sea. Based on fossil footprints found in limestone, he described what he thought must have been a voracious reptile—a bulky, fast-running predator with a strong, short neck, large head, serrated teeth, and sickle-shape claws.

Knowing his audience was absorbed, Buckland decided the time was right to dazzle his listeners even more with the details of his creature's size.[14] Cuvier had compared the ten-inch circumference of the thigh bone belonging to Buckland's beast to the thigh bone in modern lizards. Multiplying out the proportions, he estimated that it must have surpassed 40 feet in length and had a build greater than a large elephant's. Surely at this point the crowd let out a collective gasp. Buckland would have been satisfied. The meeting was turning out to be even better than he had expected.

But Buckland didn't stop there. With great gusto, he went on to describe the period of monsters in which his creature lived:

> During this period of monsters...huge lizards, their jaws like crocodiles, their bodies as big as elephants, their legs like gate-posts and mile-stones, and their tails as long and as large as the steeple of Kidlington or Long Habro. Take off the steeple of either church, lay it in a horizontal position, and place legs on it, and you will have some notion of the animal's bulk. These stories look like fables, but ask not your indulgence to believe them. There the monsters are, and I challenge your incredulity in the face of the specimens before your eyes. Disbelieve them if you can.

Buckland had no way of knowing that the *Megalosaurus* he was describing would later be deemed the first dinosaur. The muscular meat eater charged through England 165 million years ago, weighing a ton. It would be nearly another two decades—in 1842—before the great naturalist Richard Owen would put forth for the first time the name Dinosauria—which means "terrible lizards" in Latin—or dinosaur, to describe the "fearfully great reptiles" that included the *Megalosaurus*.

Sitting in the hallowed halls of the society that night was Gideon Mantell, the geologist and a country doctor from Sussex, also on the southern coast of England. His wife, Mary Ann, had recently discovered a giant tooth, this one bearing an uncanny resemblance to the smaller teeth of a modern-day iguana. Based on the tooth, Mantell conjured up a new animal, which he named *Iguanadon*. He had guessed—quite rightly—that it had been even larger than Buckland's *Megalosaurus*. In his own scientific paper, which he would present to the society six months later, in September 1824, he speculated that if the teeth bore the same relative proportions in both living and fossil animals, his *Iguanadon* must have been upward of 60 feet long.

The announcements made on that cold, dark February evening sparked a flurry of publicity and public discussion during a pivotal epoch in history. *Megalosaurus* was a major anomaly that did not square with the accepted view of the world, a fact that unsettled even Buckland and his fellow scientists. Like Buckland, many were ordained ministers. How could the *Plesiosaurus* and *Megalosaurus* make sense in the biblical equation? Recent finds were revealing the existence of animals like nothing in modern times, shoring up evidence of extinction, still an unpopular and frightening notion in the 1820s.

Buckland offered no explanation at the time. Privately he continued to agonize over the differences between his strong religious beliefs and what Mary and others were digging up from the earth. But unlike the scriptural geologists of his day, Buckland didn't hold fast to the literal interpretation of Genesis, especially in terms of geologic time. Although he rejected evolution, he acknowledged that Earth probably supported life before humans were created.

The gentlemen who attended the meeting at the Geological Society were led to imagine that the *Megalosaurus* might not have been the only giant reptile to stalk the land in ancient times. Surely there were others. They also likely struggled to make sense of—and come to terms with—the assertion that giant lizards had once roamed where they were now standing. But, they had seen the fossilized remains with their own eyes.

Back in Lyme Regis, Mary was undeterred by the theological wrangling and personal misgivings taking place in the big city, and she continued digging up fossils that fit no blueprint previously imagined. She unearthed ichthyodorulites—the fin bones that shielded the primitive *Hybodus,* the fearless ancestor of today's great white shark.[15] She discovered several new species of ammonites and continued to master the intricacies of anatomy by cutting up and studying both the soft tissue of modern fish and the dried up bones of ancient ones, without ever stepping foot inside any museum or university. In 1824, shortly after the Geological

Society meeting, she wrote to the British Museum, politely asking to be sent a full list of its collection. There also are reports that she tried to teach herself French so that she might be able to contact Cuvier in his original language.

At a time when a woman did not walk in public with a man to whom she was not related, Mary was visited frequently by many great scholars, all of them men, in search of information as much as fossils. Thomas Allan, a banker and amateur geologist from Edinburgh, visited Lyme Regis in June of that year. "She walked with us on the beach and showed us where she looked for and found her best specimens," he wrote. "She says she is indebted to her father for all the knowledge she has."[16] He told of how she had wandered, eyes fixated on the ground, and found a very fine dorsal fin of a *Hybodus,* which he later purchased. He wrote: "Mary Anning's knowledge of the subject is quite surprising—she is perfectly acquainted with the anatomy of her subjects, and her account of her disputes with Buckland, whose anatomical science she holds in contempt, was quite amusing."

Allan probably wasn't able to fathom how a poor, uneducated young woman could dare spar with an esteemed scientist. But Mary felt extremely comfortable with Buckland and had a high regard for his commitment to science; their quarrels always seemed to be innocent and even playful debates between good friends.

A short time later, an English noblewoman named Lady Harriet Silvester visited with Mary, describing her talents in this way:

> the extraordinary thing in this young woman is that she has made herself so thoroughly acquainted with the science that the moment she finds any bones she knows to what tribe they belong. She fixes the bones on frame with cement and then makes drawings and has them engraved.... It is certainly a wonderful instance of divine favor—that this poor, ignorant girl should be so blessed, for by reading and application she has arrived to that degree of knowledge as to be in the habit of writing and talking with professors and other clever men on the

subject, and they all acknowledge that she understands more of the science than anyone else in this kingdom.[17]

The fact that Mary was making this kind of impression is extraordinary considering that, at the same time, her neighbors were succumbing to the pressures of the latest superstition. When a blight, or grub worm, attacked blackberry plants near Lyme Regis in the mid-1820s, many of the town's residents got it into their heads that the destruction was being caused by some kind of flying serpent. It wasn't long before a rumor spread, claiming that this serpent also had the power to destroy anyone younger than 30. As a result, many people rushed out to buy magic charms for protection.

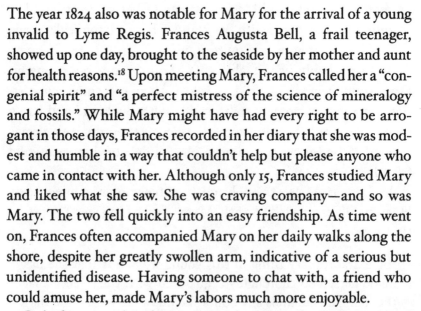

The year 1824 also was notable for Mary for the arrival of a young invalid to Lyme Regis. Frances Augusta Bell, a frail teenager, showed up one day, brought to the seaside by her mother and aunt for health reasons.[18] Upon meeting Mary, Frances called her a "congenial spirit" and "a perfect mistress of the science of mineralogy and fossils." While Mary might have had every right to be arrogant in those days, Frances recorded in her diary that she was modest and humble in a way that couldn't help but please anyone who came in contact with her. Although only 15, Frances studied Mary and liked what she saw. She was craving company—and so was Mary. The two fell quickly into an easy friendship. As time went on, Frances often accompanied Mary on her daily walks along the shore, despite her greatly swollen arm, indicative of a serious but unidentified disease. Having someone to chat with, a friend who could amuse her, made Mary's labors much more enjoyable.

Only four months after arriving, however, the Bells returned to London. Mary was heartbroken. But letters from Frances were forthcoming.

My very dear Mary, a whole month having elapsed since since I left you, and dear Lyme, without your hearing from me. I fear you will begin to think me unkind in not fulfilling my promise of writing, or suppose that illness has prevented me. We bore our journey tolerably and arrived in smoky London at nine at night. O how different from the pure air you breathe!

Then, in an intriguing reply, Mary wrote:

My dearest Fanny, Many thanks for your kind, interesting letter, and I have to beg your pardon for doubting your friendship; not hearing from you for six weeks instead of two. I thought if illness had been the cause of your silence, your dear good aunt would have sent me one line, just to tell me: the world has used me so unkindly, I fear it has made me suspicious of all mankind. I hope you will pardon me, although I do not deserve it. How I envy you your daily visits to the museum! Indeed I shall be greatly obliged your sensible account of its contents; for the little information I get from the professors is one-half unintelligible. Very little doing in the fossil world; excepting I have found a tail for baby, and a beautiful paddle, and a few other small specimens; nothing grand or new.

The afternoon of November 22, 1824, brought with it a heavy rain that raked the coastline. By evening, the pace of the rain had accelerated, pounding away at the land and the sea. Suddenly, without warning, the storm whipped up a gale that seemed intent on savaging the entire English Channel. Mary, Molly, and Joseph holed up in their cottage, where undoubtedly they were serenaded for hours by the sickening howls of the wind.

Unbeknownst to them at the time, scores of vessels were foundering up and down the shore and dozens of people were dying.[19] All through the night, hurricane-force winds pounded the shore, with 23-foot-high waves tearing apart buildings, uprooting trees, and drowning livestock. The wondrous Cobb, which had defiantly stood by the village for so long, was breached and severely damaged, as were about four dozen houses and businesses in Lyme

Regis alone. Records of horrific storms devastating the Dorset coast date back as far as A.D. 877, when over 100 Viking longships toppled in Swanage Bay. But this storm front, which stalled for several hours over Lyme Regis, was of such magnitude that it is still talked about today.

One vessel knocked ashore that night was the little trading brig *Unity*, which was sailing for London when she was caught off guard by the storm. A local resident, naval captain Charles Cowper Bennett, and two other local seamen had to run to the rescue. They managed to save those on board, despite a blinding wind and in the darkness of the night.

By 4 A.M., the Annings' own first floor was under water, and Mary was forced to move her fossils upstairs. As the day dawned, Mary surveyed an appalling landscape. Front gardens had been laid bare, trashed with shingle. The doors and windows of many homes and businesses had been torn off by the winds. At least one shop had been completely swept away.

Soon afterward, Mary wrote of the storm to Frances:

> Oh! My dear Fanny, you cannot conceive what a scene of horror we have gone through at Lyme. It is quite a miracle that the inhabitants saved their lives. Every bit of the walk from the Assembly rooms to the Cobb is gone and all the back parts of the houses on the shore side of town.... All the coal cellars and coals being gone and the Cobb so shattered that no vessel will be safe here, we are obliged to sit without fires this winter, a cold prospect you will allow.

For Mary, the storm was a significant event. She had never before witnessed such destruction—and it happened just when she was truly coming into her own. An article in the *New Monthly Magazine*, a popular literary journal at the time, had referred to Mary as "the well-known fossilist, whose labors lately have enriched the British Museum as well as the private collections of many geologists."[20] Finding the plesiosaur had firmly sealed Mary's reputation as a first-rate fossil hunter, even though gentlemen scientists had failed to give her her due.

And, undoubtedly, Buckland and De la Beche were still trying to get over the fact that Mary had outfoxed the famed Cuvier, who had been forced to acknowledge that her skeleton wasn't a forgery. But now the town around her had been decimated right before her eyes.

6

The Hidden Mysteries of Coprolites

In the last weeks of 1824, the skies stayed mostly clear. However, this wasn't enough to reassure the storm-weary Lyme residents, many of whom were determined to get as far from the reach of the sea as possible. Even Mary, who loved the ocean, and her mother were eager for a change. As the months passed, however, the reminders of the weather's violent outburst were assuaged by the many visitors with only sun and sea on their minds who returned to the town the following spring and summer.

Soon there were so many visitors in Lyme Regis that dozens were unable to secure accommodation and had no choice but to seek rooms in other coastal towns. In response, a new crop of entrepreneur rose up to meet the demand: More and more residents opened boardinghouses, rented out rooms, and began building full-fledged hotels. While famine stalked much of the country in the 1820s, Lyme Regis continued to cash in on its ability to charm tourists. A few who had traveled abroad liked to compare Lyme Regis favorably to a Turkish village, at least from afar, from the intervention—or placement—of trees, plants, and gardens, and the irregularity of the houses.[1] If nothing else, Lyme Regis remained a stalwart safety valve, able to accommodate guests when visitors to Bath overflowed.

Throughout the rest of Dorset, though, there was acute depression. Small industries that had prospered during the Napoleonic Wars suffered from a loss of orders in the years following the war's end in 1815. Some 400,000 soldiers had returned home, dramatically swelling the ranks of the unemployed. Heavy rains and particularly cold winters during this time also served to hamper agricultural production.

Nevertheless, both Bath and Lyme Regis continued to profit from what seemed to be almost weekly proclamations from researchers that sea air was a major contributor to longevity. Even for those not scientifically inclined, the fact that 17 people had lived beyond 80 and another 29 had made it to nearly 90 in a town of about 2,000 was all the proof anyone needed of the coastal area's advantages.[2]

By the summer of 1825, with dozens of men hard at work repairing the Cobb—even today a masterpiece of masonry several hundred feet in length—it was almost as if the storm had been a long-ago nightmare. According to George Roberts, the town shrugged off the destruction and seamlessly went back to being a place of genteel resort, "where good society is enjoyed in a fuller extent at a much cheaper rate than in any town of this description." He boasted that the town offered "the fairest prospect of restored health and spirits to the invalid."[3]

For every indulgence available in big cities like London, Paris, and Rome, Lyme Regis offered simple pleasures—picnics and boat rides to the romantic cliffs—as well as unbridled morality. During the early 1800s, the community apparently was so bent on ensuring that its members adhered to certain high standards that it was not above reprimanding its own. A "skimmington ride" was a sentence that could take many forms, but mostly it was a combination of pain and shame at the hands of one's neighbors, swiftly dealt to any individual viewed as being sinful or disruptive.[4] The concept of "skimmington"—or taking the law into one's own hands—was viewed in France and elsewhere on the Continent as a facet of low-brow culture. But almost all the residents of Lyme, regardless of their social status, seemed to have abided by this widely held form of rough justice. When a woman talked back to her husband, or a husband struck his wife, or a couple fought too loudly, a posse of community members might suddenly turn up on the offender's doorstep, clanging pots and pans and making raucous music on whatever "instruments" they could round up. Sometimes an effigy was constructed. More often than not, the offender was carried through the streets astride a pole supported by two stout lads. Driving home the humiliation even more, demonstrations typically were repeated over several nights.

In addition to justice, superstitions continued to thrive in Lyme Regis during this time. It wouldn't have been uncommon, for example, for residents to tack old horseshoes above their doors, supposedly to fend off evil spirits. And few would have dared to eject a toad from their cellars without the greatest of care, so associated were these animals with prosperity. Chimneys, too, were paid special attention, with pieces of bacon suspended inside, a particularly bizarre practice designed to interrupt the descent of witches.

For the most part, the people of Lyme Regis were kind and just, usually struggling to stay financially afloat with both religion and irrational superstitions featuring prominently in their lives.

Mary Anning believed her discoveries and scientific method were beyond reproach and she would have put this down to her own intelligence and perseverance, not to any superstitions or good luck charms. She was never one to blink owlishly into the spotlight. Rather she was infected with an abundance of confidence. Her good friend Anna Maria Pinney once wrote that Mary was afraid of no one and even intimated that Mary was discontented with her own people.

> She says she stands still, and the world flows by her in a stream, that she likes observing it and discovering the different characters which compose it. But in discovering these characters, she takes most violent likes and dislikes...Associating and being

courted by those above her, she frankly owns that the society of her own rank is become distasteful to her, but yet she is very kind and good to all her own relations, and what money she gets by collecting fossils, goes to them or to anyone else who wants it.[5]

It must have irked Mary greatly that her gentlemen scientist friends never had to worry about money, while she scrimped and saved every bit she earned.

In 1826, when Mary was just 27, she had saved enough money to purchase—not rent—a small but attractive cottage farther from the sea on upper Broad Street, just down the road from the Philpot sisters.[6] Leaving behind the crowded Cockmoile Square, the family moved their belongings into the back of the house while Mary turned the front room into a proper shop with a glass window so that passersby could see inside—just the type of shop her father had dreamed of. Drawing attention to it was a modest oblong white sign that read: "Anning's Fossil Depot." In Lyme Regis, moving to higher ground also implied a moving up in social status. Mary must have been enormously pleased that she was able to take such a big step, especially as economic troubles were handcuffing so many others across the country.

They moved to the part of Lyme Regis known as Top of the Hill, a desirable section, with cows grazing nearby, that felt a lot more rural than their former home. George Roberts had moved to the neighborhood as well, and lived right across the street. A few blocks downhill was the newly opened Royal Lion Hotel, a fine establishment where many of Mary's fossil customers stayed over the years. The lower part of Broad Street was the town's most important shopping district, always the site of constant construction and reconstruction. Around the corner from Broad was Sherborne Lane, a nearly vertical, century-old cobblestone walkway that spiraled down to the river Lym, hemmed in by thatched cottages on both sides.

Mary's move was so momentous it was even reported in the *Sherborne Mercury* newspaper on July 22, 1826: "Miss Mary Anning's Fossil Depot is removed to Broad Street where it will always prove a source of attraction." The article even told readers that there was a "splendid animal" currently on display, an "Ichthosaurus Tenuirostris." Above all else, owning their own home rather than paying rent would help the Annings weather the tough economic times to come.

Mary's reputation as a talented fossil collector not only led to a more comfortable living arrangement but also to a rush of visitors. For them, the long journeys to meet with Mary would have been an exercise in tenacity, as coach trips were interminably slow, especially after a good rainfall, which could turn the roads of southern England into a muddy quagmire. Always a little stir-crazy in closed-off Lyme Regis, Mary appeared to have welcomed each caller with enthusiasm.

Perhaps the person who went to the greatest extremes to visit Mary was the geologist George William Featherstonhaugh, who journeyed all the way from America in October 1827 to collect fossils for the recently opened New York Lyceum of Natural History.[7] Naturally he sought out Mary. Apparently he appeared out of the blue, introducing himself and saying that he was keen to purchase her finest specimens. As always, Mary would have been only too happy to oblige.

Born in 1780, Featherstonhaugh was a graduate of Oxford University who traveled in 1806 to the United States, where he fell in love with and married an American woman. By all accounts, this trip to England was a fortuitous one, and he proved to be engaging company for Mary. Later he declared her to be a "very clever, funny creature." His visit led to a plethora of items, many purchased from Mary, that were shipped from England for the New York museum and also to more orders for Mary from collectors in the United States.

When it came to Mary's personal life, another encounter, this one with Roderick Murchison, an English geologist of great stature, paid the biggest dividends. Whereas Mary had been an early bloomer scientifically, Murchison was a late starter—and seldom has one who started so late been able to achieve so much so quickly. The son of Scottish gentry, he joined the army at 15, resigned after the defeat of Napoleon, and lived and breathed only for hunting until, prodded mercilessly by his forceful yet seemingly irresistible wife, Charlotte, he finally agreed to move to London to pursue science instead of the fox. At the advanced age of 33, he took up geological fieldwork and went on to become one of the last independently wealthy gentlemen geologists of the era.

By the 1820s, it was becoming widely accepted by most everyone, churchgoer or not, that the successive rock layers were laid out on top of one another in such a fashion that the lower rocks were the first formed and the upper rocks the latest formed. Five geological "systems" were established. Murchison's claim to fame was the discovery of a new strata while traveling in South Wales, a trip made at Buckland's urging, to study the fossiliferous rocks underlying a rock formation known as the Old Red Sandstone. This new strata contained fossils of marine invertebrates that predated the appearance of land plants and vertebrates. Murchison's naming of a Silurian time period—443 million to 417 million years ago—won him acclaim both at home and abroad. Eventually he presided over the Geological Society, the Geographical Society, and the British Association for the Advancement of Science before being knighted in 1846.

But before any of this, Roderick and Charlotte Murchison's first geological adventure was along the coast of southwest England—and they never forgot it. After a short visit in Lyme Regis, Roderick thought it best to leave his wife behind for a few weeks so that she might benefit from the sea air and "become a good practical fossilist, by working with the celebrated Mary Anning of that place, and trudging with her, pattens on their feet, along the shore."[8] Mary probably always had made good use of pattens, a design that included a flat metal circle touching the ground, a metal plate nailed into the shoe's wooden sole, and metal bars connecting the two, all of which helped Mary keep her feet out of the mud.

Together, Mary and Charlotte spent countless hours combing the coastline for vestiges of the long dead. Only 11 years older than Mary, Charlotte Murchison was everything Mary was not: a poised, well-groomed, more mature woman of good breeding with impeccable manners. But, like Mary, she had a taste for adventure, and wasn't above getting dirty.

After that first visit, Mary corresponded with Charlotte Murchison for many years; in one letter, Mary described Charlotte's husband as "certainly the handsomest piece of flesh and blood I ever saw."[9]

With Mary as her guide and inspiration, Charlotte Murchison spent the rest of her life searching for fossils, both with her husband and on her own. She identified and labeled specimens, sketched fossils and the rock formations where they were uncovered, and became a lifelong fan of Mary's. Charlotte was a dynamic, energetic, and sagacious woman who held great sway over her husband. She was also fiercely loyal. Any time she encountered a collector during her travels, she never failed to mention Mary's fossils, thus helping Mary build up a remarkable network of contacts across Europe.

In 1828, shortly after their visit to Lyme Regis, the Murchisons embarked on a long journey across Europe with Charles Lyell, the young Scottish lawyer and another of the era's leading geologists. Thanks to an efficient, if lopsided, division of labor, the men decided the routes and research topics while Charlotte was left with much of the grunt work. Not only was she charged with sketching the landscapes and geological structures—on top of numerous hours of fossil collecting—but she also was the who conversed with local experts, as she was the only one in the party fluent in French. Without Charlotte, the trip would have been far less successful.

But while the Murchisons presented a real picture of a partnership that was for the most part loving, tender, and successful, Mary's good friend Henry De la Beche was finding himself in a disastrous relationship that had finally ripped apart completely.

In the mid-1820s, De la Beche had journeyed to Jamaica to spend a year on his family's plantation while carrying out a full geological survey of the eastern part of the island.[10] Upon his return to England, he poured out his findings in various papers that presented geological data on the island—at the time a very remote part of the world to most Europeans. He also published a controversial pamphlet entitled *Notes of the Present Condition of Negroes* in which he sought to "state fairly and candidly" what he himself had witnessed.

By all accounts a decent businessman, as well as a kind person, De la Beche had worked hard to acquaint himself with his own slaves, going to great lengths to ascertain whether they were content. After an informal investigation, he ordered the overseers in his own fields not to carry whips but to exact punishments at a later time, presumably after everyone had calmed down. Although abolitionists were gaining more support and slavery was on its way out, De la Beche always maintained that his own slaves had been well cared for. It was obvious that without the service of slaves, plantations would become less profitable, and indeed the profits did start diminishing year after year. Within the next decade or so, De la Beche would be on the brink of losing his inheritance, facing the prospect of having to actually work for a living.

This was a challenge De la Beche wouldn't have been afraid to meet. But adding to the sting was the fact that his personal life was in shambles. While he was away in Jamaica, his wife, Letitia, took a lover. In 1825, she asked for a legal separation because "the union proved to be of the most unhappy nature: the treatment which Lady De la Beche received at the hands of her husband being such as to render it impossible for her to live with him." De la Beche admitted "hasty expressions" and added that he "ought to have trusted entirely [her] own high sense of honor in many cases."

Because of the fame De la Beche had won for his geology work both at home and in the Caribbean, Letitia's affair and the marriage's subsequent breakup was a public scandal, creating an especially taxing period in the life of Mary's friend.

Perhaps De la Beche was too caught up in geology to understand what went wrong. Details of the split are sketchy, but he must have wondered if it was his time in Jamaica that had derailed the relationship. In those days, however, it was common for many couples to spend lengthy periods apart.

Their official separation came in 1826, after which Letitia moved in with her lover, Major General Henry Wyndham, son of the Earl of Egremont. In a highly unusual move, the courts granted De la Beche custody of his two daughters, but by then his health was apparently so poor due to the acrimonious nature of the whole messy affair that he opted to allow the children to remain with Letitia. He wrote: "Misfortune has followed me from my cradle, and it will follow me to my grave." In despair, he chose to go abroad in an effort to take his mind off his troubles. The decision was a smart one. He spent months crisscrossing France and Switzerland, and it was his research in these countries that later consolidated his position as one of the leading geologists of Britain.

Mary might have secretly harbored hopes of forming her own relationship with De la Beche after the breakup of his marriage. Many authors and historians have hinted at a romantic tie between the two. Perhaps she had imagined that the close bond the two had forged in their teens—built on a foundation of innumerable days perusing the cliffs together—might one day have blossomed into something more. In those times, even an independent woman would have yearned to be married, especially to someone able to lift her up in status. But De la Beche may have left England for Europe after his breakup without a word to Mary.

Fortunately, around this time Mary found a welcome distraction: her old friend William Buckland, who was becoming increasingly obsessed with what would become known as coprology—the study of fossilized feces. When he showed up at her door during his Christmas holidays in 1826, she was surely delighted to see him.

Only a year earlier, Buckland, in his early 40s, had married Mary Morland, a 28-year-old collector of fossils from Oxfordshire with a passion for geology. Already an accomplished artist, she had worked closely with no less than the eminent Cuvier, illustrating some of his finds as well as the works of Buckland's friend Conybeare. Buckland couldn't have asked for a more supportive spouse or congenial companion. She assisted him without complaint in everything he did, even taking on the laborious task of labeling every one of his fossils, a job that would have kept her up night after night. Reflecting their mutual penchant for fossils, both agreed to spend their long honeymoon touring geological sites in Italy, France, Austria, Germany, and Switzerland. Clearly married life agreed with Buckland, who went on to have nine children, five of whom survived to adulthood.

After exchanging pleasantries, Buckland would have explained to Mary that he was in Lyme Regis for business. In particular, he had grown intrigued by some of her smaller finds. For years, Mary had stumbled across twisted, rounded dark-gray pebbles, some with black spots, while out searching for fossils. Often they were no more than four inches long and only an inch or so in diameter. But sometimes they were much larger. The same type of object seemed to occur in a variety of shapes including cones, spirals, and cylinders. For as long as Mary could remember, the locals had called these masses "bezoar stones," since they were shaped like the hard gallstones belonging to the ubiquitous bezoar goat, the ancestor of the domestic goat. The more superstitious among the locals believed these bezoar stones contained protective medicinal properties.

On at least a few occasions, Mary had found these stones inside the skeletons of ichthyosaurs, leading her to believe they might be fossilized clumps of undigested food that remained in the intestines, if not ejected at death. She showed Buckland how she had found many of these clumps within the abdomens or pelvises, or in very close proximity to the skeletons. To her, the reason these masses had taken on their puzzling shapes and spiral markings might have been obvious: They simply had passed in a soft form through the intestines of ancient animals. Initially, though, Buckland wasn't convinced; he believed that the stones were molded around bits of clay and that they were probably formed fairly recently. He was baffled as to how these ubiquitous masses fit in with the prehistoric world. By the time he visited Mary, he wanted to know more.

Buckland had studied excrement in the past, notably the petrified feces of hyenas in caves located in Kirkdale, in northern England. Discovered by workmen in 1820, these ancient animal dens were packed with the bones of hyenas as well as the species they devoured. Oddly shaped clumps found there were made up of calcium phosphate and carbonate and often contained fragments of bones or plant residue.

After hearing Mary's take on the stones, Buckland consulted Gideon Mantell, the struggling doctor and geologist, who had come across similar stones with marked whorls in the sandstone of the Tilgate Forest, located between Lyme Regis and London. Comparing pages of notes, both agreed that the bezoar stones were found mostly in strata that was fertile with animal fossils. Upon further investigation, Mantell decided that Mary's conclusions were right on the mark: The stones were indeed fossilized feces. A gleeful Buckland now had everything he needed to determine what the ancient beasts had eaten. In 1828 he named the fossilized feces "coprolites," from the Greek *kopros*, meaning "dung," and *lithos,* meaning "stone."

The study of fossilized feces would turn out to be one of the most important techniques available to paleontologists. But initially it was hampered by its very nature—one that was deemed indelicate within at least some geological circles in Georgian England. Certainly it wasn't a study for those easily nauseated.

Mary and Buckland, though, jumped into the study of coprology with their usual zeal. In order to properly examine the coprolites, the pair first had to rejuvenate the hardened feces into a soft malleable substance, probably working in Mary's home. Reconstituting the feces also resurrected their overpowering former stench. But there was no other way to determine exactly what was inside the feces unless the masses were softened. And there was plenty inside, all of which yielded meaningful clues as to the habits of the early creatures. In Buckland's view, they had hit the jackpot. "When we see the body of an Ichthyosaurus still containing the food it had eaten just before its death," he said, "ten thousand, or more than ten thousand times ten thousand years ago, all these vast intervals seem annihilated, come together, disappear, and we are almost brought into as immediate contact with events of immeasurably distant periods as with the affairs of yesterday."

Picking through the feces, they found remnants of fish, fragments of bone, as well as undigested teeth and vertebrae, which led them to conclude that the more powerful creatures had preyed on the weaker ones. Buckland described it as "warfare waged on successive generations of inhabitants of our planet on one another; and the general law of nature which bids all to eat and be eaten."[11] Mary once had happened upon two coprolites near one another and close to the skeleton of an ichthyosaur, "as if they had been voided by it in the struggles of death." Years later, it was coprology that led scientists to determine that the *Tyrannosaurus rex* was a carnivore.

Buckland was so intoxicated by what coprolites were able to tell him that he even had a new dining room table made, inlaid with polished coprolites. At the same time, in an era when "toilet humor" was not uncommon, a few of Buckland's students also had a field day with his new area of study. One of them, John Duncan, was inspired enough to write this poem:

> Approach, approach, ingenuous youth,
> And learn the fundamental truth
> The noble science of Geology
> Is bottomed firmly on Coprology.[12]

Poring through the fossilized feces surely would have provided ample opportunity for the two enthusiasts to get to know one another even better. Most likely, they engaged in some wonderful conversations. It might have been during those days that Mary first learned of her friend's efforts to reconcile geology with the Bible. At this time, Buckland was still profoundly frustrated that he'd been unable to find concrete evidence to support important biblical stories, let alone evidence of a worldwide flood. Likely he admitted that he was now acting under the assumption that God may not, in fact, have created the heavens and the earth in the first day; the words "in the beginning" might instead refer to an epoch of time. But now, in light of what coprology was telling him, Buckland also was growing increasingly concerned over why such a kind God chose to fill the primitive world with such seemingly evil carnivorous beasts.

Buckland would spend years trying to come to grips with the idea of a benevolent God having a hand in violence. To him there could be only one good reason for it: God had wanted to give every creature the ability to make the end of life for every other creature as straightforward as possible. And the easiest of deaths is proverbially that which is the least expected and so sudden that it doesn't leave a lot of time for suffering. To some, it might have seemed like convoluted reasoning, but apparently few were able to come up with a better explanation. The only thing Mary would have known for sure, after scrutinizing her own fair share of fossilized feces, was that the ancient creatures must have been aggressive hunters inclined toward eating anything smaller than themselves.

The progress they were making—humble woman and wealthy professor—was unprecedented. The promises of coprology tantalized everyone with an interest in prehistoric life. Until this time, scientists were never sure exactly how living things turned into fossils. But if something as pliant as undigested food could be preserved, it was plausible that some fossils were created by the sudden entombment of animals before even the softest tissues had had time to decompose. Usually, when something dies in nature, it's either scavenged or decayed by bacteria. But a different process occurs when something becomes fossilized. Creatures such as the ichthyosaur likely were covered in water and mud after dying. Because the body was cut off from air, the normal process of decomposition did not play out in exactly the same way as it would above ground. Very often, bodies settle on the ocean floor. Although some of the fleshy parts might be eaten or rot away, the skeleton generally remains intact and eventually is covered in layers of sediments. Over the years, these sediments seal up bones and just about anything else—including undigested food. In this way, Mary's discovery of coprolites led scientists to a new understanding of how fossils were formed.

Unfortunately, though, no one had ever become rich studying feces. After Buckland traipsed back to Oxford, Mary again faced the task of trying to sell fossils. The late 1820s were just as bleak as the early part of the decade. By this time, many of Mary's most reliable patrons were suffering their own financial hardships. The Duke of Buckingham, for example, who earlier had paid good money for several of the Annings' fossils, was in such dire circumstances by the mid-1820s that he could no longer bid for a fossil. So embarrassed was he by his burgeoning debts that he actually moved out of England for two years.

Another customer Mary typically could have counted on in the past, the Bristol Institution, also was facing a financial implosion after several local banks collapsed following damaging speculation in the London market. In all, some 60 banks in London and Bristol crashed in 1825 and 1826, setting off a nationwide panic. Added to that was the repressive effect of the Corn Laws—import tariffs designed to protect British farming—which caused erratic fluctuations in food prices. With even few public museums buying specimens, the prices of fossils began to plunge.

For a while, Mary felt that her only choice was to concentrate on finding smaller fossils that tourists would be willing to buy. She and Elizabeth Philpot began spending an inordinate amount of time in the grip of what Mary dubbed "the green sand mania," scoping out the higher—and newer—layer of nearby cliffs known as Greensand in an effort to find tinier curiosities. They invested countless hours "beating bits of green sand to pieces to find shells."[13] But their efforts mostly proved fruitless, and soon Mary decided to give up and go back to the Blue Lias she knew so well.

In a stroke of luck, Mary happened upon another discovery in 1826—one that benefited all the people of Lyme Regis. One day, after cutting open one of the long bullet-shape fossils called belemnites, Mary noticed a tiny chamber containing what looked to be dried-up ink. She showed it to Elizabeth Philpot, who added some water to it, creating a dark liquid concoction similar to a painter's sepia. Elizabeth dipped a brush into the brownish solution and began drawing with it. Soon she was using the ink to complete all sorts of artwork. She even finished one of an ichthyosaur and sent it off as a gift for Mary Buckland in a sweet gesture of friendship. Mary's discovery of the "paint" was a boon for tourism as local artists began churning out drawings of prehistoric beasts sketched with the ink found inside other prehistoric creatures.[14]

Over the next year of so, Mary sold William Buckland a number of the belemnites she found with the ink sacs inside, some nearly a foot long. Studying this prehistoric ink, Buckland also could see what Mary probably had recognized long before any scientist: These ancient creatures had hid from predators by squirting a cloud of ink out into the water around them.[15]

The summer of 1828 gave way to fall. Though she'd scarcely missed a day on the shore, Mary had very little to show for her work that year. She had uncovered a fine specimen of *Dapedium politum,* a ray-finned fish about 10 inches wide that lived about 160 million years ago, but little else. She probably put on a brave face for her worried mother, assuring her that she would find something substantial soon, but perhaps Molly wasn't so sure. Maybe, they both would have feared, all the good stuff had already been found.

Mary's heart may have been heavy around this time. De la Beche was off once again on another tour of the Continent. Buckland was back at Oxford. In her spare time, she pored over Lord Byron's poetry and transcribed whatever scientific articles she could get her hands on. She desperately yearned to make another great fossil find.

In 1828, Lyme Regis's Blue Lias was being quarried in greater and greater quantities to meet the demand in London and other cities for stucco. External stucco had been introduced into the capital city toward the end of the eighteenth century and grew increasingly popular thanks to an appetite for smooth, evenly colored house fronts. The destructive quarrying of limestone began to open up larger areas of Blue Lias, accelerating the disintegration of the cliffs. The tides wore them down even further and unmasked even more fossils. The reward, for Mary, was a larger search area.

On a brutally cold but bright clear morning in early December, Mary may have been in no mood to go to the beach. But she went anyway, just as she always did. After only a brief time outside, she probably felt cold and wind-whipped. As always, Tray stood dutifully nearby, his eyes fixed on Mary's every move. Mary's pickax had battled against many a rock when suddenly she caught a glimpse of something peculiar in the cliff side. The sight would have made the hairs stand up on the back of her neck.

Usually when she found a fossil, it was like a cog in the wheels of her imagination. One look at a bone, and she could almost conjure up an image of an ichthyosaur or a plesiosaur swooshing through the water. But she had no idea what this particular hodgepodge of fossils might be. After a swift perusal, she would have been stumped. To her delight, the winter storms, coupled with the recent flurry of quarrying and her digging, had revealed a winged creature unlike anything she had ever encountered.

In contrast to the giant sea lizards Mary had uncovered in the past, this ghostly relic was slight and misshapen, with a long tail bursting with dozens of vertebrae and culminating in a diamond-shape tip. There were hints of claws and wings and an enormous skull memorable for its rounded jaw and long beak. After a bit more digging, all was revealed: The entire fossil was less than four feet long. Although not the greatest creature she'd ever uncovered when it came to size, it certainly was one of the most unearthly ones; it looked to be a cross between a vampire bat and some kind of reptile. Indeed, the pièce de résistance was its three shorter clawlike fingers followed by a fourth outlandishly long finger.

Fragments of hollow birdlike bones had been discovered before at Lyme Regis, but no firm identification had ever been made. When Buckland heard of Mary's find, he hurried to see the bones for himself. He had long suspected that such a flying creature might have existed, especially after finding some disarticulated bones of a wing and a toe while browsing through the Philpots' collection. But the idea of a giant flying lizard had seemed fantastic and implausible. Now, with Mary's find, his suspicions had been confirmed. He declared: "Miss Mary Anning...has recently found the skeleton of an unknown species of that most rare and curious of all reptiles."[16]

An image of Lyme Bay from 1823 by an unknown artist. Photo courtesy of Dorset Coast Digital Archive.

An illustration of Broad Street, Lyme Regis's main commercial street, in 1845 attributed to C. Wanklyn. Photo courtesy of Dorset Coast Digital Archive.

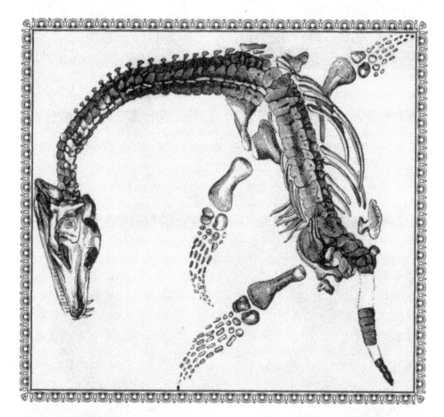

An image of the *Plesiosaurus*, by William Buckland, one of Mary's major finds. Wellcome Library, London.

Portrait of the eccentric Professor William Buckland, a longtime friend of Mary's and fellow geologist. Wellcome Library, London.

An image of Lyme Regis's shoreline from 1841 attributed to C. Wanklyn. Photo courtesy of Dorset Coast Digital Archive.

Duria antiquior, a depiction of Mary's finds, drawn by Mary's good friend Henry De la Beche in 1830.

Henry De la Beche: A childhood friend, De la Beche was a highly respected geologist and Mary's lifelong supporter.

Engraving showing Lyme Regis from Church Cliffs published by F. Dunster. Photo courtesy of Dorset Coast Digital Archive.

An image of Lyme Regis's shoreline from 1841. Image courtesy of Dorset Coast Digital Archive.

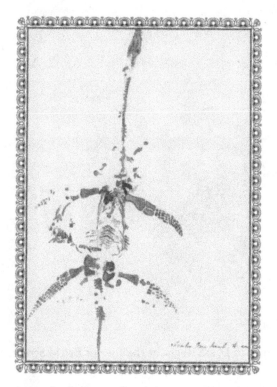

An early plesiosaur fossil discovered by Mary.
Wellcome Library, London.

Painting of Mary Anning and her dog Tray by an
unknown artist. Today it hangs in the Natural
History Museum in London.

The Fossil Depot in Lyme Regis, at the foot of Broad Street, one of many fossil shops that came after Mary's death. Photo courtesy of Dorset Coast Digital Archive.

Engraving of Professor William Buckland giving a speech to the Geological Society at Oxford University by N. Whitlock. Wellcome Library, London.

7

Finally, the Big City of London

$\mathcal{F}$ew relationships were as great a source of solace as Mary's relationship with the Philpot sisters. The friendship was fostered by many days outside on the beach as well as many days inside the Philpot home, one of which may have been similar to the following.

Darkness had barely left the early-morning sky on one of the last days of February 1829 when Mary charged out for her morning constitutional, headed, as she often was, for the Philpots' home. We have no evidence of what the home looked like, but it might have had pretty sash windows and rosewood-veneered furnishings, with a handsomely designed living room. It might even have felt far too fancy for Mary to sit in. Most likely Mary and Elizabeth and at least one or two of Elizabeth's other sisters would have migrated to the dining room, where walls were hidden behind several glass-topped cases with shallow drawers down the front, chockful of fossils with little tickets attached, scribbled with information about when and where each item was found.

As always, Mary would have been struck not only by the gracious splendor of the rooms, richly decorated in silk and velvet, but also by the courtesy with which she was treated. Although Mary was closest to Elizabeth, who was 19 years her senior, she apparently got on with all the sisters, whose impeccable manners would have been a sharp contrast with those of the brassy sailors she sometimes encountered on the shore, and even with those of Mary's lower-class contemporaries in town. As the ladies sat down, perhaps with dainty cups of tea before them, the atmosphere might have been electric. After all, it had been only two months since Mary's discovery of the strange winged skeleton. But it was around this time that Mary learned that her dear friend Frances Bell had died, succumbing to a disease that was never recorded.

Upon hearing the news, Mary had rushed off a note to Frances's mother, explaining how much she had appreciated Frances, not only as a friend but also as an inspirational spiritual guide. She wanted her to know that "the recollections of Frances' pious conversations have been a support to me in the trials I have had to sustain; this has enabled me to say 'Not my will, but thine, O Lord, be done.' My dear Madam, we must endeavor to look on this world as a state of trial, to fit us for a better life."[1]

Mary and Frances had exchanged many letters over the years, often laced with religious overtones. Indeed, Mary's one comfort was that Frances' soul surely had risen to that place where there would be no more pain. For Mary, corresponding with Frances might have been like corresponding with a sister, one who understood the confines of Lyme Regis but who could also have acted as her eyes and ears in the big city. Without a husband and children, Mary had few emotional outlets; she would sorely miss the comforting messages and updates she and Frances had exchanged.

Once, after the havoc wreaked by what was now referred to as the "Great Storm" of 1824, Frances had soothed Mary with these words: "let us hope that He who permits nothing to happen but for wise and beneficent, though perhaps by us unperceived, purposes, will repair your losses in good time. I assure you, dear Mary, that in all my afflictions I have trusted in the same firm Rock, and as yet I have never been deceived."

As she aged, the impetuous fossil hunter was becoming a deeply meditative woman who often could be found praying or reading the Bible. Ironically, it was around this time that her most recent fossil treasure, the flying reptile, was raising even more doubts about the validity of parts of the book both Frances and Mary knew and loved so well.

Elizabeth would have sympathized with her friend and undoubtedly reassured her by pointing out how deeply Frances regarded Mary's companionship and correspondence. Surely, Elizabeth would have told her, Frances would have been proud of Mary's success. Most likely Mary regretted that Frances had died before hearing of her most recent discovery.

Eventually the conversation at the Philpots' home would have turned to the new skeleton. Mary might have complained about how little she'd heard from Buckland regarding what the learned gentlemen were making of the find, beyond the fact that it inescapably underscored the magnitude of reptilian life that must still be buried within Lyme Regis's terrain. As women were not allowed at meetings of the Geological Society, Buckland had had the honor of describing Mary's creature in lecture and in writing, within the *Transactions of the Geological Society* for February 1829. Although the eccentric professor, always a loyal friend, made note that it was actually Mary who had found the skeleton, officially he was given credit for the discovery. Privately, this probably made Mary exceedingly resentful.

But even Mary would have admitted that only Buckland had the oratorical talents to bring to life her discovery in such dramatic fashion. Indeed, his description would have stuck with the members for a good time to come:

> It somewhat resembled our modern bats and vampires, but had its beak elongated like the bill of a woodcock, and armed with teeth like the snout of a crocodile; its vertebrae, ribs, pelvis, legs, and feet, resembled those of a lizard; its three anterior fingers terminated in long hooked claws, like that on the fore-finger of a bat; and over its body was a covering...of scaly armor like that of an iguana; in short, a monster resembling nothing that has ever been seen or heard-of upon earth, excepting the dragons of romance and heraldry.[2]

It wasn't long before news of the skeleton reached Paris, where Charles Lyell announced, albeit with less flash and flair than Buckland, Mary's discovery before a class of students being taught by French geologist Louis-Constant Prevost. Only five years earlier, Prevost and Lyell had visited Anning, marveling at her discovery of a two-foot ichthyosaur. Both men knew her work well.[3]

Mary's discovery would eventually be called a *Dimorphodon,* the earliest type of Jurassic pterosaur. It was the first pterosaur—or "winged lizard"—ever discovered outside of Germany. Like all pterosaurs, Mary's batch of bones initially looked something like a toy model of an animal whose parts had been mismatched by a careless child. For one thing, the head appeared to be far too big for the body. Also surreal were the two types of teeth in its jaw— massively long ones out front, presumably for snatching prey, and shorter sharp ones in back, presumably for grinding up what they caught. Indeed, the name *Dimorphodon* means "two forms of teeth."

But what really set the pterosaur apart from other early reptiles were its wings, the structure of which was formed by a slick leathery membrane that stretched like elastic between its body, the top of its legs, and its elongated fourth fingers. It was believed that by flapping their wings, the creatures could fly with relative ease, even though the largest ones probably were forced to rely on updrafts and strong winds to help them with liftoff. First appearing about 200 million years ago—almost 70 million years before the first known bird—pterosaurs had existed alongside dinosaurs.

A contemporary of Mary's earlier finds, the ichthyosaur and the plesiosaur, the pterosaur was believed to be the biggest creature ever to take to the air. Over time, other pterosaur skeletons would be found. Although some were as small as today's birds, others had wingspans of nearly 40 feet. Like their dinosaur cousins, pterosaurs lived on Earth for millions of years before dying out, leaving no trace of an explanation for their disappearance.

Again, though, Mary wasn't the first to find remains of the birdlike pterosaur. That honor went to Italian zoologist Cosimo Collini, a former secretary to Voltaire. Collini discovered his bones in a limestone quarry near Solnhofen in northern Bavaria in 1784. Initially he guessed he was staring at a kind of sea creature—perhaps an amphibious mammal of some sort. But later, in the early nineteenth century, Georges Cuvier reviewed illustrations of the remains and quickly interpreted them as being from a flying reptile—a notion promptly dismissed as being too far-fetched.

It was Cuvier who coined the name "Ptero-dactyle," which means "winged finger," for that first specimen uncovered in Germany; however, due to the standardization of scientific names, the official name for this species morphed into *Pteroactylus*. Still, the name "pterodactyl" and "pterosaur" became interchangeable over the years and eventually both stuck, popularly used to refer to all members of this specimen's order.

For the Philpots and for Mary, the pterosaur was just one more mystery in an ever-shifting, enigmatic universe. After discussing the skeleton that morning, Mary might have recognized that the sisters were as confused as she was. She would have wondered whether the gentlemen were making better sense of things than they were.

The year 1829—when the first steam-powered carriage traveled from London to Bath at a speed of 15 miles per hour—was a momentous one not only for Mary but also for the country doctor who devoted his life to studying prehistoric creatures. In 1829 Gideon Mantell released the seminal paper that is still talked about today: "Age of Reptiles." This paper was the first to set out detailed evidence of an era of reptiles that preceded an era of dominance by mammals, thereby suggesting a slow progression from primitive reptiles to advanced mammals a quarter century before Charles Darwin published *Origin of Species.*

Mantell's take on Earth's history, 245 million to 65 million years ago, was thought by many to be a wild-eyed one, vividly encapsulating the terrifying idea of a time in which "the earth was peopled by quadrupeds of a most appalling magnitude. These reptiles were the Lords of Creation before the existence of the human race!"[4] Although Cuvier himself had acknowledged that reptiles came before mammals in the development of vertebrates, he had never been able to turn this tacit recognition into any sort of coherent argument. Mantell hoped his paper would spawn widespread interest that would result in financial support. It did not.

To his dismay, Mantell's theory drew mostly shock and scorn from biblical literalists, and almost no monetary rewards. Most likely his assertions were simply far too radical—and far too nightmarish—for the time. By the late 1820s, diabolical beasts from the past were gripping the public's imagination like never before, with some wondering whether snaky bodies might still be alive, lurking beneath the leaden cloak of the sea. These fears were given a boost by Mantell's paper, which portrayed certain animals as a separate creation, independent of man, in a world where a human being wouldn't have been safe anywhere. The image defied the sensibilities of many, with educated critics and even the humblest farmers unable to fathom such a thing as an Age of Reptiles.

One of the loudest outcries against Mantell's theories came from the Reverend William Kirby, a highly respected scholar who absolutely refused to accept that such creatures as pterosaurs were extinct. Surely, he reasoned, they still must be flying about somewhere, breaching only occasionally into sight in unexplored areas of the world. His was a belief shared by many. Even Mary probably sometimes wondered whether she might turn a corner on the Lyme Regis shore, only to catch a sudden glimpse of the bulging eyes of an ichthyosaur or the crushing teeth of a pterodactyl.

The opinions of Kirby, considered the father of entomology, were thought to be sacrosanct. It was Kirby's beautifully written, four-volume handbook, *An Introduction to Entomology,* that helped spark a craze in England for collecting beetles. And, in America, one of the most widely read books was Kirby's painfully titled *On the Power, Wisdom, and Goodness of God as Manifested in the Creation of Animals and in their History, Habits, and Instincts*. The book included fastidious descriptions of animals and animal migrations and closed with these words: "What can more strongly mark design, and the intention of an all-powerful, all wise, and beneficent Being, than that such a variety of animals should be so organized." And so Kirby—who vociferously rejected any challenge to God's purpose—could not have been more aghast by Mantell's hypothesis.

"Who can think that a being of unbounded power, wisdom, and goodness should create a world merely for the habitation of a race of monsters, without a single rational being in it to glorify and serve him!" he railed.

After announcing Mary's find and describing it to the world, Buckland found himself riding a wave of popularity in England. His steadfast defense of the idea of a worldwide flood was finally starting to pay off, with the notion becoming an article of faith for many members of the Geological Society. There were still attacks on his assertions, but they generally took the form of humorous disparagement. In early 1829, a popular rhyme that poked fun at him was passed around by students and professors alike: "Some doubts were once expressed about the flood, Buckland arose, and all was clear—as mud!"[5] But in general, Buckland's positions in the church and at Oxford gave him the utmost authority on matters of both science and religion.

An enthusiasm for the flood theory of geology swept not only through Britain but through much of Europe by the late 1820s. The idea that water had once engulfed the planet, as laid out by the Bible, was a comforting one, and so clear that fewer and fewer skeptics dared to counter it. Buckland was pleased so many people supported his viewpoint. While conceding some challenges in efforts to harmonize geology and Christianity, he continued to argue that there were incontrovertible truths that simply could not be ignored. One needed only look at the stratified formations of peaks and valleys, or at the confluence of streams, to see undeniable proof that Earth was once covered by water.

Support for Buckland's conviction stretched all the way across the Atlantic, where, in America, one of the more vocal supporters of a universal flood was Eleazar Lord, the president and founder of the New York and Erie Railroad. In his book *The Epoch of Creation,* he went much further than Buckland by charging that it was impossible to know the secrets of the world's origin except through divine revelation. He attacked those who sought to forge a link between modern geology and the Bible, and he bitterly denied that a perfect God would muck about for millions of years trying to improve an imperfect creation. Like Buckland, he argued that the existence of a flood was the only way to make sense of the geological record.

But there was one person who was not afraid to raise doubts about the flood viewpoint: Charles Lyell. Indeed, though he was a Christian, in the late 1820s Lyell became more determined than ever to "free the science from Moses." England, he declared, "was more parson-ridden than any country in Europe except Spain."[6] And this, he charged, was detrimental to scientific thinking. But when his seminal work, *Principles of Geology,* was finally published in 1829, he was careful not to overtly attack the Bible or the Genesis narrative.[7] Instead he chose to release a cautious but heavily researched account of the main discoveries that he and others had made up to that point—and to let them speak for themselves. He succinctly spelled out the evidence for why he believed there had been radical changes in climate and geography across the ages, and he speculated on the slow but progressive development of life that resulted. He raised the idea that valleys and mountains were formed not by the biblical flood but by erosion from local rivers. And he argued that countless minute changes had fashioned Earth's crust over vast periods of time—all according to established natural laws.

Lyell's interpretation of Earth's history as being vast and directionless greatly influenced the younger scientists of the day. Their impressionable minds were intrigued by his conclusion that, because the processes that alter Earth are uniform through time, "huge Iguanodons might reappear in the woods, and the ichthyosaurs in the sea, while pterodactyle might flit again through umbrageous groves of tree ferns."

One man who devoured Lyell's landmark text was Charles Darwin, who read it while aboard the HMS *Beagle,* a British Royal Navy ship that sailed in 1831 on a five-year expedition from one exotic locale to another. Later Darwin confessed to being much inspired by the work, galvanized in particular by Lyell's strong yet subtle ridiculing of both the belief in a worldwide flood and the belief in a recent creation rather than an ancient one. Darwin's own experiences during his voyage only backed up the geologist's theories about how the shifting crust of Earth was one of the ongoing, gradual, planet-sculpting forces.

But in America, Benjamin Silliman, the country's most influential man of science in the early nineteenth century and Yale's professor of chemistry from 1802 to 1853, was making his own arguments that science works only to reveal "the thoughts of God." Today the proposition that the universe displays a creator's intelligent design has provoked bitter debate throughout the United States. But in the early 1800s, no one in America dared deny intelligent design. Even the leading critics of organized religion insisted that the physical universe displayed evidence of providential interference. Indeed, they liked to contrast the firm manifestation of God's existence supplied by nature with what they considered the weaker proofs provided by the Bible's "fables."

This back and forth—often quite heated—energized the churches, the universities, and, ultimately, the dinner-table conversations of average people.

While these debates on its origins and significance raged on around the world, Mary's pterosaur was also firing up the public's imagination and fueling people's fascination with geology and paleontology—a fascination so great that national museums were having a hard time keeping up with the demand to view fossils. To meet this burgeoning interest, entrepreneurial collectors starting setting up their own private exhibitions, proudly laying out their natural trophies in their own homes, particularly in large cities.

Until the Natural History Museum was completed in London in 1881, the city's rapidly mushrooming collection of natural specimens—mostly donated by Sir Hans Sloane—was stored in the Natural History Section of the British Museum. The British Museum, the first in Europe explicitly owned by and open to the public, is today the world's oldest scientific academy in continuous existence, attracting visitors from around the world. But in those days, the Royal Society was the dominant force, controlling not only what went on display at museums but also people's access to them, thus influencing the way in which the natural world was introduced to, and examined by, the public. Certainly the museum helped to consolidate the legitimacy of scientific pursuits, but the process always took place under the Royal Society's watchful eye.

Although she had never been inside a museum, Mary was becoming intimately acquainted with the country's museums through her written communications. In particular, she had been in direct and frequent contact with the officers of the Bristol Institution for the Advancement of Science, Literature and the Arts since its inception in 1823. Sealing the relationship was the fact that the first geological specimen acquired by the museum had been Mary's fine skeleton of an *Ichthyosaurus communis*. It had been donated by the group of scientists that included De la Beche and Conybeare. She sold another exquisite ichthyosaur skeleton to the museum in the late 1820s for £50, a sizable sum at the time. Unlike the Bristol Institution, the British Museum had been so financially strapped it had been unable to scrape together the funds to buy Mary's *Pterodactylus*. Buckland wound up buying the creature with his own money to prevent the nation from being embarrassed by the piece being sold to a foreign institution.

In the United States, the public's growing interest in the natural world spawned the founding of the Academy of Natural Sciences in Philadelphia, the oldest natural sciences institution in the Western Hemisphere, which opened its doors in 1828. The museum housed vast collections of plant and animal species, including many gathered by Lewis and Clark during their exploration of the American West. In time, scientists on this museum's payroll trekked everywhere, from the Arctic tundra to the Amazon, to study and catalog forms of life both living and extinct. As in Britain, an interest in nature was also in vogue among the upper classes in the United States during the early 1800s. Many people displayed their large collections at home. Indeed, the Philadelphia museum's important collections were made possible in large part due to the gifts of wealthy collectors.

Against this backdrop of public interest, it's no wonder the beaches of Lyme Regis were pulsing with new blood as budding geologists tried their luck at beachcombing and perhaps even trying to catch a glimpse of the illustrious fossil hunter Mary Anning. The upper classes were motivated by the combination of travel and romantic engagement with nature. The middle classes who followed in their footsteps sought to advance socially.

The year of 1829 was a serendipitous one for Mary, bringing about good fortune and good luck. In March, she uncovered the second complete skeleton of the long-necked plesiosaur. The *Dorset County Chronicle* reported that

> Miss Mary Anning of Lyme has discovered another specimen of the Plesiosaurus Dolichoderius. This specimen, which is 11 feet in length, is almost perfect, and most of the bones are lying in perfect order. The head, sternum, vertebrae, and bones of the pelvis and the paddles are all fine and in place. There are four vertebrae between the last dorsal vertebrae and caroidal vertebrae, with their false ribs attached; by which it appears that this creature had the power of shifting its sternum, property of

some amphibious animals now existing, extremely curious, but very useful when swimming.[8]

So magnificent was the skeleton that an international battle erupted between museums wishing to be the first to showcase it. Initially the Academy of Natural Sciences in Philadelphia expressed a desire to purchase it, sending Buckland into another fury over the idea that England might lose one of its own natural treasures. Immediately he demanded that the British Museum buy the skeleton, which it finally did, after much prodding, for £105 pounds. After the sale, accolades praising Mary's prowess poured in.

The *Edinburgh Philosophical Journal* also made mention of her fossil finds that year, along with those of the Philpot sisters, Charlotte Murchison, George Cumberland, and Gideon Mantell, in the first-ever published list of Britain's greatest geological collections. That same year, Cumberland made note in the *Royal Institution of Great Britain's Quarterly Journal of Literature, Science and the Arts* that Mary's "industry and skill" had led her to find "nearly all the fine specimens" of ichthyosaurs ever discovered.[9] Even the venerated Charles Lyell gave a nod to Mary's talents by writing to beg her assistance in determining whether the sea was impinging on the shoreline east and west of Lyme Regis.

On the downside, Mary was now 30—a milestone that entrenched her in the status of old maid. That same year, 1829, her older brother, Joseph, now 33, married a woman named Amelia Reader and moved to a home on nearby St. Michael's Street. From the moment they had met, there had been inexplicable tensions between Mary and Amelia. Although Joseph apparently continued to look after his mother and sister as best he could, even after he married, Mary may have resented that he was no longer as fully involved in the family's fossil business. Now that he had a wife to support, perhaps Joseph would want to devote himself full time to his more reliable employment as an upholsterer. But Mary might not have seen it that way. She also might have been jealous that her only sibling—her constant companion since birth—had found someone to share his life with while she hadn't.

Although Mary had participated in and helped shape the history of geology, the scientific elite remained hostile to intellectual women in the early nineteenth century. Home and family were the distinctive duties of women, and, as Charles Dickens noted in his writings years later, there were signs that even the people of Lyme Regis looked askance at Mary's pursuits. Also hampering Mary's advancement was the difficulty faced by any woman wanting to do work away from home. Everything from a lack of chaperones to inadequate funding made it hard for women to travel much beyond their immediate environs.

Undoubtedly Mary resented that church appointments—with their prestige, security, and financial rewards—were enabling her scholarly gentlemen friends, like Buckland, to pursue their passions in relative comfort while she was forever struggling to make her next great find merely to stave off hunger. She had risen from obscurity but had also grown weary of living vicariously through the exploits of her friends. Time and again, De la Beche had spilled tales of his travels to her as they wandered the beach during his visits to Lyme. His latest was a tour of Italy, during which he studied the rocks around the gulf of La Spezia near Genoa. The trip was not all work. He visited the Holy City of Rome—which he described as a place where "personal vanity could be easily indulged"—and even sat for a cameo portrait there.[10]

Ever since she had been a girl, Mary had longed to see the world, even once lamenting to Charlotte Murchison how very sorry she was that she had not been able to accept an earlier invitation to visit the Murchisons in London. "I have never been out of the smoke of Lyme," she complained in a letter.[11]

But by the summer of 1829, it may have seemed to Mary that anything was possible—even traveling far from Lyme Regis. And so in July she finally took the Murchisons up on their open invitation. She was going to London. She so looked forward to the journey that she commented "should anything occur to prevent my accepting [the invitation], it will be the death of me."

Mary loaded a trunk filled with clothes and fossils onto a ship that departed from the Cobb every other Saturday, bound for the capital. The ship, the 80-ton *Unity*, was the same one that carried her fossils from Lyme to the scientists who studied them. Under brilliantly sunny skies, the vessel slowly navigated along the southern coast and then traveled up the busy Thames River, a days-long but exhilarating ride for someone who had never before traveled. Finally Mary would have landed at London Bridge, disembarking amid a frenzy of forces and faces.

At this time, fewer than two-thirds of London's nearly 1.7 million inhabitants were actually born in the city. In the mix were probably blacks, Chinese, Indians, Poles, Frenchmen, and Italians. Everyone seemed to be in a hurry. At the same time, the air was thick with the smut and dank smells of a coal-fired city. Every one of Mary's senses would have been overcome. Everywhere she turned, there were people. Indeed, by the late 1820s, the city's infrastructure was cracking under the strain. Improvements were ongoing, but slow to be completed.

For example, for six centuries, London Bridge had been the only pathway across the Thames; in 1750, a second stone bridge called Westminster Bridge was erected to meet the demands of a burgeoning population. The old London Bridge was approaching the end of its life, and Mary would have taken note of the rise of a new London Bridge by designer John Rennie that was, during her visit, nearing completion, right next to the old one. Eventually, two years later, it would be opened.

Mary was struck by the multitude of people and their movements.[12] The capital's sheer scale was astonishing, especially to someone from the country. London was not just big; it was incredibly hectic, with crowds and clatter everywhere. And it was also sordid, plagued in parts by black rats, which had been driven out of the sewers by their stronger rivals, brown rats. To critics, London had become iniquity personified, a hothouse of vice surrounded by pestilential slums. Compared with Lyme Regis, which was smoky in parts but still widely renowned for its fresh sea air, the smoke-filled atmosphere of London was extraordinary. Coal fires burned in almost every house. The resultant smoke stung Mary's eyes and clung to her clothes.

Nevertheless, Mary was invigorated by her time in the nation's capital. London's diversity was thrilling, almost like a garden bursting with flowers she'd never encountered. Walking the streets was like witnessing an inexhaustible theater of sights and sounds. Not only was London the world's largest city at this time, but it also was the world's greatest manufacturing center, its greatest port, and its greatest financial center. London was at the pinnacle of its importance, a bustling metropolis teeming with immigrants lured by the possibility of employment.

For someone who had never spent a night away from Lyme Regis, she would have been awed by the Murchisons' home at 3 Bryanstone Street, just south of Regent's Park, a house probably full of footmen, fine art, and deep crimson colors. Also grand was the park itself, a 450-acre near circle.

Elizabeth Philpot might have lent Mary a nice lacy dress for her adventure, but the brawny young woman, with her tanned skin and strong arms, still would have stood out against the pale faces and delicate physiques of the upper-class women strolling in the Murchisons' neighborhood. In those days, women wore their hair parted in the middle and styled elaborately over the temples and down the sides. The wealthier women donned wide-brimmed hats garnished with masses of ribbons and feathers, and tight corsets that made even breathing difficult at times. Mary, in contrast, generally made her own clothes, all plain and loose. And she usually wore her hair on top of her head, under a hat or bonnet. As a Dissenter, she would have been taught that any excess in dress and manner was offensive, so she was likely taken aback by the elaborate outfits worn by women who looked as if they'd never done a day's work in their lives.

Mary's tour of London naturally started at the Geological Society located at the magnificent Somerset House, a recently constructed neoclassical gem overlooking the Thames. Since the middle of the eighteenth century there had been a rising outcry over the lack of great public buildings in London. As a result, Somerset House had been constructed in 1775, looking more like a grand palace than public offices. For Mary, it was intoxicating to visit the scene where Conybeare had first described her plesiosaur. She was led through the building by geologist William Lonsdale. He guided her through the zoological rooms, pointing out every manner of stuffed bird and animal. He apparently went to great lengths to explain the gradations between the mastodon, elephant, and rhinoceros, calling her attention in particular to a model of a jaw of a large *Megalosaurus*.

Next on her itinerary was a visit to the British Museum, where she was especially enamored with the King's Library—donated by George III in 1823—that housed some 60,000 volumes. She also enjoyed seeing her first diorama—in this case, of Rome. These three-dimensional models of mostly urban developments had been invented in 1822 and were popular with tourists.

It was only fitting that Mary squeezed in a visit to Sowerby's. George Brettingham Sowerby had been her agent since 1824. He and his son, both conchologists and artists, sold a variety of natural history specimens, including fossils and minerals. Charlotte Murchison, forever looking out for Mary's best interests, had initially put the two in contact. Sowerby's was situated on King Street, which fed into Covent Garden Market, with its special bouquet of colors, smells, and sounds. It, too, was changing. Between 1828 and 1831, a new market building went up, after which the buzz surrounding the place could attract an even more fashionable clientele. Mary likely saw wagons heaving with cabbages amid outdoor stalls buried under shiny fruit, while rosy-cheeked Irish girls peddled baskets full of peas, beans, and turnips.

Mary was even more fascinated by the sophisticated shops up and down Regent Street, a fashionable district designed by John Nash, the famous architect behind the reconstruction of Buckingham Palace in 1825. Named after the prince regent, later George IV, the street was built as part of a grand ceremonial route from his townhouse to the villa George planned for Regent's Park.

Mary had never seen anything like the grandeur of Nash's terraced houses surrounding the park, with their many columns and pediments. From her own notes on the visit, she found London to be a dazzling paradox of poverty and prosperity, a spectacle of the changing times. She was thrilled to finally be in the thick of things. But keeping up with the fast pace must have taken a mental toll.

With its inequities, crime also thrived in London during the 1820s. Thousands of brothels, gin shops, and gambling dens flourished with few checks on their activities. The same year as Mary's visit, a bill establishing the city's first real police force was passed by Parliament. Behind its passage was the steely resolve of future British prime minister Robert Peel, who was keen to use the publicity surrounding the new force to further his political career. Almost no one could deny that the old system of watchmen wasn't working, despite frequent public hangings. But the new police—called "bobbies" after their founder—were in for a rough initiation. They were vilified, spat on, stoned, and generally molested by a public that viewed the force as a threat to British liberty. Protesters marched through the city streets, smashing windows, starting fires, and attacking all those in uniform.[13]

Mary was used to living in a mostly safe environment, one in which nature posed more of a threat than any unsavory characters. She wouldn't have been completely naive to the ways of the world, though. Lyme's harbor had made the town fairly cosmopolitan with French-, Russian-, and Scandinavian-flagged ships and their foreign sailors calling frequently. Mary also had been introduced to dark faces, usually those of servants aboard ships from Lyme involved in the triangular trade: England, to Africa, to America, to England.

Mary returned to Lyme Regis with tales of lung-choking smog and high prices, of fancy dresses and music-filled churches. She was delighted by the Elgin Marbles but may have missed the rhythm of the waves pounding on the Lyme shore.

At the end of such an adventure, Mary would have been in fine form, at the pinnacle of success. Throughout the year, she had accomplished much, and the timing of her finds was still proving to be impeccable. Everything was clicking.

8

An Amazing New Fish

The autumn months of 1829 were full of rain, and more rain. And it wasn't just drizzling rain, but the kind that makes everyone take cover. The dreadful weather might have matched Mary's mood. For a woman who had recently achieved some semblance of success—and who had just been immersed in the exhilarating hustle and bustle of London life—sitting on the sidelines again would have made for an uneasy perch. In Lyme Regis, only her lonely and quiet work on the craggily coast awaited her. By this time, the shoreline might have looked a lot more hollow and forlorn than it used to. Although only 30, Mary might have felt much older than her age, after so many years of physically taxing labor. She might have looked a lot older as well. But in spirit she still would have been young enough to have gotten sucked in by the undeniable allure of London's crowded streets and alleyways.

Nevertheless, she went back to her fossil hunting with the kind of resolve that descends on racers near the finish line. As always, taking stock of herself, she would have harbored hopes of fossil hunting her way into greater financial independence—to her, money represented freedom. And so she plunged back into her coastal world of clays and mudstones not only for escape but, again, out of necessity. She continued to exhibit a real patience with—and aptitude for—the work, and the exercising of her faculties gave her a sense of value, as it had for so many years. No matter how she was feeling, Mary would crawl out of bed early, almost every day, whatever the weather, and coax an enormous amount of work from her weary body. She paced back and forth across the quiet confines of the shore with an air of purpose, squarely focused on the task at hand as she carefully sidestepped rock pools and perilously slippery formations, slowed only by the mucky ground and frigid winds.

Throughout the long, darkening days of October and November, Mary was disappointed to have come up with little to show for her risky hunting efforts. In an unsentimental letter to Charlotte Murchison, she described one recent incident. It was an afternoon during which Mary found herself stalled for a good while by a bone she felt sure belonged to some kind of plesiosaur. She had asked a quarry worker to stand by, in case she needed any help, while she poked and prodded, on and on. But her stubborn attempts were interrupted by the sudden onslaught of a treacherous tide, one that seemed to form out of nowhere.

No matter how well Mary knew the coast, an offshore storm system still had the ability to strengthen the tide before she knew what hit her. And prevailing winds could change course swiftly, charging through the rock and sand, forming a powerful channel of water that could flow back out to sea in a dangerous rush. Fortunately, Mary was a fast runner, and she and the quarry worker managed to make a narrow escape. By the time they reached her house, though, they looked like a "couple of drowned rats."[1] She asked the man why he'd never warned her about the tide. He admitted that he had noticed it and that he had been frightened, but since Mary, a woman, appeared not to be bothered by it, he had been too ashamed to call out in warning. It was just another reminder of how precarious fossil hunting could be, especially in the colder months, when new specimens were more likely to be exposed because of the wildness of the weather.

In a cheerful footnote, Mary apologized to Charlotte Murchison for the poor quality of her letter, in which several words had been misspelled, saying: "I beg to say that I have walked ten miles today and am so tired I can scarcely hold the pen."[2] Although she corresponded with some of the most influential minds in Britain, and indeed in Europe, Mary still had little use for punctuation—even when not tired—and had her own manner of spelling. Often she came up with excuses for her hit-or-miss grammar. It was a reminder that she was, after all, a young woman with almost no formal education.

More weeks went by and still she wasn't finding anything even remotely worth pursuing, the exception being that one plesiosaur bone, which hadn't led to anything. Her feelings of impatience would have mounted. But finally her persistence paid off, as it always seemed to—and in spectacular fashion. It was the afternoon of one of the first really clear days of early December 1829, a month that always seemed to be golden for Mary. As always, the sheer cliffs would have been patterned with abstract shapes created by the limestone and softer clay, geological artwork that was typically blue gray in color due to its heavy iron content. Suddenly, off in the distance, she must have noticed a flash of white against an otherwise dark backdrop. She almost hadn't spotted it.

As soon as she chipped away some of the mud surrounding the bone, buried near a crevice slightly off to one side of the menacing Black Ven, she had a hunch she was on to something. After so much rain, the fossil's tomb of overlying rock slipped away easily with very little prodding. One bone led to another and then another and then another, all part of a fossil that wasn't enormous in size but certainly was peculiar in appearance. A mere 18 inches long, it had a long snout and looked a bit like a fish, but not a regular one. Mary gathered up whatever bones she could before the tide rushed in, so eager to show them to her mother that she would have run if the ground hadn't been so typically sticky.

The next morning, she was rejuvenated and on the cliff, pickax and shovel in hand. This time she toiled for hours while at least a handful of onlookers staked out the scene. At this point, more than a few of the residents of Lyme Regis fancied themselves to be amateur fossil collectors, although they were no match for Mary's years of experience. Mary wasted no time in exposing the entire skeleton—all but the tail, that is. It was nowhere to be found. Everyone could see that this fossil was smaller than her previous major finds, but still, it was odd enough to cause a considerable stir. Fortunately, the bone fragments were easy to pack and transport, and not so brittle as to fall apart when touched. After Mary got them home, her brother and the Philpot sisters would have dashed over to have a look.

And so Mary had found another fossil that was about to make a splash, and the *Salisbury & Winchester Journal* was among the first to report the find to the public. Its description was straightforward: "It is about a foot and a half long; has two immense sockets for the eyes, a long snout, a series of the finest vertebrae that have ever been seen in so small a creature, claws, fins like wings."[3] Mary's meticulous restoration had brought to life a beautiful specimen. Dissecting a modern ray, she was quite confident her fossil was a fish from an entirely new species of ray. In bed at night, Mary must have quietly thanked God for allowing her to make another major discovery.

Mary was able to foresee only one hitch: trying to sell the reconstructed fossil in an economic environment that was continuing to deteriorate. Battling the financial storm, even the wealthiest of collectors were eager to horde cash, reluctant to part with their money for such a luxury as a fossil. Revolutions were breaking out across Europe, and there were more and more riots in England over the price of bread, which was the staff of life for the lower classes. These financial difficulties bore bad tidings for the Annings. But there was good news too. By this time, Mary had become an astute saleswoman who understood that her find's novelty could still help her secure a decent price.

Immediately she launched into action, carefully drafting two letters bearing news of her fossil. In the first, to Charles Lyell, Mary initially referred to the find as "nondescript" but then went on to make sure he was aware of her inability to associate it with any other known creature. Mary's second letter was to Buckland. And he was the first to respond. Conscious of her continuing precarious financial predicament, he urged her to expand her marketing efforts by reaching out not only to London-based collectors but also to the Bristol Institution, where her close friend De la Beche was an active member. Mary followed Buckland's advice, and while waiting to hear back from Bristol, she summoned up the courage to also write to Adam Sedgwick, the Cambridge professor who had taught geology to Charles Darwin, with details of the specimen.[4]

As part of her sales pitch to Sedgwick, she included a sketch of the fossil of professional quality, which she modestly called "a scratch." She described her find as "a skeleton with a head like a pair of scissors," a unique fossil, "analogous to nothing." Sedgwick wasn't interested, and neither was Lyell. Eventually, though, a wealthy landowner named John Naish Sanders came forward with an offer. Eighteen months after she'd nearly given up trying to sell it, Sanders purchased Mary's fossil for the Bristol Institution, paying about £40. Although Mary promised to send the tail when she found it, the Philpots stumbled upon it first, and so this appendage wound up in the Philpot collection.

Once again, another of Mary's finds was sparking intense interest among the big guns of the Geological Society in London, who this time were divided over whether the fossil was a reptile or a bird or something else altogether. After Mary sold the fossil to Sanders, Dr. Henry Riley studied it at its new home in the Bristol Institution. At his fingertips at the institution were all the tools of the early nineteenth century comparative anatomy, and he took full advantage of every one of them.[5]

Only two years older than Mary, Riley was a local surgeon and medical school teacher who had been trained in Paris. He also was a member of the important group of gentlemen naturalists who founded the Bristol Institution in the 1820s. So fervent was his desire to know all there was to know about anatomy that Riley wasn't above robbing graves, and he was arrested at least once for doing so. His obsession made him the perfect person to first describe Mary's fossil fish scientifically.

Although some scientists had figured it to be a reptile or bird, Riley issued a different verdict: It was a fish. More specifically, it was a cartilaginous fish that was very much like rays in some ways but very different at the same time, boasting all sorts of inimitable characteristics. Riley did not have to convince Mary of its rarity. She had verified this for herself by her own dissection of a ray. But Riley's analysis wasn't flawless; he confused the snout and the frontal spine with the lower and the upper jaws. This led Professor Robert Grant, a highly respected London anatomist, to raise questions about the legitimacy of Riley's assertions during one of his lectures on fish anatomy. As Grant had not inspected the fossil for himself, the public suspicions surely raised Riley's hackles.

Amazingly, it would take another four years, until 1833, before scientists agreed that Riley had been right all along and that the fossil was indeed a fish. Yet some continued to argue the meaning of its various attributes. Debates raged on—for decades. Eventually, though, Mary's fossil was named *Squaloraja polyspondyla,* a fish-eating chimaeroid with a body like an otter's and a flat tail like a beaver's. With such a mishmash of peculiar features, it's not surprising the name *chimaeroid* was derived from a fire-breathing she-monster in Greek mythology that boasts a lion's body and a serpent's tail.

Riley had made a smart call about an unfamiliar specimen. Mary's fossil was indeed that of a cartilaginous fish—a fish later lumped together with the likes of sharks, skates, rays, and other vertebrates with internal skeletons made entirely of cartilage. More specifically, the fish—swimming the seas some 150 million years ago during the time of the late Jurassic Era—was ancestor to both the shark and the ray. To the untrained eye, the *Squaloraja*'s importance might have seemed somewhat contrived. But it was a significant find in that the creature was able to carve out its own niche at a time when the all-powerful dinosaur had reigned supreme. What's more, *Squaloraja* was another extraordinary transitional creature, this time between sharks and rays. As such, it provided yet another piece of proof for the "chain of being" viewpoint that saw all of nature's wonders as cogs in the same wheel, links that were ultimately controlled by the same omnipotent creator.

Most important for Mary, it was her fourth major discovery, one that kept her in the spotlight—for the time being.

On June 26, 1830, only a few months after Mary found her *Squaloraja,* King George IV died suddenly at Windsor Castle, just minutes after grabbing his page's hand and exclaiming in a loud booming voice "Good God, what is this? My boy, this is death."[6]

The self-absorbed king's indulgent lifestyle had finally gotten the better of him. Not only was George extremely overweight, but he also was addicted to both alcohol and laudanum. He suffered from a litany of ailments: gout, dropsy, cataracts, breathlessness, and arteriosclerosis. He also began exhibiting flashes of insanity, often disquieting his stunned subjects with beguiling but completely untruthful tales about his days as a soldier who fought gallantly at the Battle of Waterloo. In his later years, his obesity, gluttony, and heavy drinking increasingly made him a magnet for ridicule, at least on the rare occasions when he allowed himself to be seen in public. He became more and more reclusive, eventually spending interminable stretches of time in bed.

But early in his reign he had generated an increased interest in seaside travel. As a king who took matters of taste, style, and leisure more seriously perhaps than his royal duties, he longed to indulge his childhood passion of going to the shore. For this he turned to Brighton, a city he grew to love so much that he transformed it into a favorite haunt for London's elite. As Prince of Wales, George IV had first visited Brighton in 1783, at aged 21, partly on the advice of doctors who thought that the sea water might ease the swellings in the glands of his neck. From his first visit, he took up the idea of making the small city a seaside spa, commissioning John Nash to build a Brighton Pavilion in the style of a fantastical shoreline palace inspired loosely by the Taj Mahal. The rich flooded in, followed by the middle classes. As a result, the town developed quickly, with grand squares laid out alongside classical terraced houses. But in the end Brighton wound up being more vulgar than charming with the *New Brighton Guide* of 1796 advertising it as a place "where the sinews of morality are so happily relaxed that a bawd and a baroness may snore in the same tenement."[7] It was this anything-goes atmosphere that George reveled in the most.

During his time on the throne, the king had done nothing to improve the lives of the Annings or other impoverished Britons. So despised was King George that even the establishment newspaper, the *Times* of London, published an unflattering obituary upon his death: "There never was an individual less regretted by his fellow creatures than this deceased king. What eye has wept for him? What heart has heaved one throb of unmercenary sorrow? ... If he ever had a friend—a devoted friend in any rank of life—we protest that the name of him or her never reached us."

George was buried in 1830 in Windsor Castle and, because he had no children, was succeeded by his brother, William IV. Although William reigned for only seven years, from 1830 to 1837, the more congenial and pragmatic king was highly regarded for his efforts to improve the lives of the poor by introducing various reform measures, among them an education act, kinder poor relief laws, and the abolition of slavery.

Yet, during the years of his reign, there was a real unease growing across England amid the contours of a new industrial era. Desperation and discontent remained nearly universal in agricultural boroughs such as Dorset, particularly following years of reductions in the amount of relief available for the poor. Under William, these mounting anxieties finally came to a head, morphing into a heroic age of popular radicalism. Violent protests broke out in town after town, especially among those who had returned from the Napoleonic Wars, only to find themselves unable to earn more than embarrassingly low agricultural wages, if they were able to find work at all.

During the reign of George and William's father, King George III, in the eighteenth century, only one class in Britain—the aristocracy—held sway in both politics and economics. The rest of society had been broken up into small pockets built on status rankings, which were not the same as social classes. Social classes were created around economic interests while status rankings were founded on appraisals of the honor of one's occupation or family lineage. All this changed by the 1830s, though, when the terms *working class* and *middle class* became very much a part of the vernacular, reflecting the mushrooming of social tensions. Indeed, the developing social classes were turning into little more than conflict groups held together by the subjugation of poorer people.

When William took the throne, his top priority was not only to help the poor but also to restore the tarnished reputation of the crown he had inherited. At one point during his rule he flattered the American ambassador at a dinner by histrionically announcing that he regretted not being born a free, independent American, so high was his regard for the nation that had given birth to George Washington, who William believed to be the greatest man who ever lived. Exuding charm, he helped mend an Anglo-American relationship that had been sorely damaged by his predecessor, not to mention the War of 1812. At home, William presided over improvements in transportation. Steam engines began to be used not just for ships but also for carriages. Indeed, railroads were built for trains of carriages and vans that were hauled by a single steam engine. The first modern railroad, the Liverpool and Manchester Railway, was opened the very first year of William's rule, in 1830. Soon the whole country would be crisscrossed by a network of railways that allowed people and goods to be transported much quicker than ever was dreamed possible before.

With high taxes, poor wages, and economic uncertainty all conspiring against her, Mary was finding it even harder to sell fossils in the 1830s than in the 1820s. She was hard-pressed to find even one wealthy gentleman willing to pay for such frivolities. The best museums, too, were wary of spending limited funds for even the finest and most complete of skeletons.

Everywhere she looked, people were drifting down the road to despondency. They fretted about everything—and so Mary would have fretted too. Lying awake at night, she might have worried about what financial calamity would befall her next. Surely she would have felt she'd come too far to find herself despondent yet again. Out of sheer necessity, she was forced during this time to rely more on the services of agents, such as George Sowerby in London, who were able to build a network of contacts in a way she wasn't able to do from Lyme Regis. But these alliances cost her dearly. Typically, these middlemen pocketed 10 percent of what the purchaser paid as the commission for arranging a sale, a cost that cut deeply into Mary's takings. But by the 1830s, fossil selling was becoming so daunting a challenge that Mary agreed to pay commissions of 20 percent—and sometimes even as much as 30 percent. She became so dismayed that she once confessed to Sowerby that she was thinking of giving up fossil hunting for good.[8]

But Mary was never the type to quit. Instead, she simply resolved to sharpen her marketing skills. Because she knew there was a bigger pool of buyers for smaller specimens than for larger ones, she again honed in on more modest finds. For example, she collected countless ammonites and brittle stars because she knew tourists had a weakness for them. She also started corresponding more with her customers, at the urging of her mother. If nothing else, a letter acted as a reminder that Mary was still out there, toiling away, in obscurity. Even after years of practice, writing so many letters wouldn't have been easy. Mary had attended school for perhaps two or three years, at most. And the Dorset dialect she spoke, along with her thick West Country accent, might have made her hard to understand when she talked with educated gentlemen from the cities. This was probably the case when she composed letters to them as well. Illustrating her mettle, though, she posted one letter after another to potential buyers, each containing an intense sales pitch.

In spite of Mary's best efforts, big sales were few and far between, and again the Annings were finding themselves in financial difficulty. But as was often the case, just when her situation started looking most bleak, Mary was tossed a lifeline by a man. This time it was her old friend Henry De la Beche who came to the rescue. Never failing to keep up with news of Mary and her work, even when crisscrossing the Continent, this companion since youth presented Mary with a gift in 1830—perhaps a belated birthday present—one that she would never forget.

Although there is no firm account as to what happened, De la Beche must have rolled into Lyme some weeks after Mary's birthday, which was on May 21. As always, he would have been a welcome sight, presenting himself at the doorstep of Mary's house, looking smart in his pantaloons and high-waisted jacket, with a kerchief around his neck. While she fetched him a cup of tea, he unpacked something from a large piece of brown wrapping paper. He gestured toward it when she returned to the room, a sly smile widening across his face. For a few minutes, she wasn't sure exactly what she was looking at. But then it would have become clear. Most likely, the image took her breath away.

The present was a watercolor entitled "Duria Antiquior—A More Ancient Dorset."⁹ It portrayed every one of her finds—three types of ichthyosaur, a plesiosaur, a *Dimorphodon*—all graphically displayed. Mary probably delighted in the fact that De la Beche had included the distasteful emission of coprolites from a terrified-looking plesiosaur. The painting accurately portrayed magnificent discoveries, springing to life in what was the first graphic depiction of the carnage and brutality of the prehistoric world. The focal point of the drawing, a large toothy ichthyosaur, was shown biting on the slender neck of a plesiosaur, while another plesiosaur was lunging from the shallow water in a sneak attack on a crocodile basking on the shore. In the distance was the elongated neck of a plesiosaur hurling itself into the air, trying to take a nip out of a low-flying pterosaur. Indeed, wildly snapping jaws infested all parts of De la Beche's image.

When she asked why he'd painted it, De la Beche no doubt explained that the depiction was designed to pull her out of her financial slump.

Fully aware that fossil prices were falling, De la Beche, an amateur artist since he had been a young man, had taken it upon himself to do something to generate a flow of income for the Annings. The plan was an ingenious one. George Scharf, one of the best illustrators in London, would turn De la Beche's drawings into lithographic prints. Copies would then be sold with the proceeds passed on to Mary. Later, enlarged copies of the drawing would be used as props by professors such as Buckland, thus sparking a new round of interest in Mary's fossils. Indeed, De la Beche's overriding goal was to entice more people to buy her specimens.[10]

De la Beche's scene was a remarkable achievement if for no other reason than because of the unique perspective it offered spectators. Portraying the panorama from the viewpoint of the marine creatures themselves, De la Beche made observers feel as if they were actually in the water with the creatures—and this was two decades before the invention of the aquarium. Later, during the Victorian age, it became de rigueur among the upper classes to display aquatic samples in glass containers in drawing rooms. Ultimately, De la Beche's idea of prehistoric life had an enormous impact on how artists illustrated scientific books in the future.

Although he depicted a violent scene, the watercolor was far from being apocalyptic. The images were far too cheerful for that. De la Beche was nothing if not a man with a keen sense of humor. Once, knowing how obsessed Buckland had become with coprolites, he produced a lithograph showing the distinguished Oxford professor of geology, decked out in formal robes and top hat, standing at the entrance of a cave that was shaped like a cathedral while brandishing about a geological hammer as if conducting a choir. Every member of that choir was portrayed in the unseemly act of defecating—including Buckland himself.

With the income from the lithographs produced from De la Beche's painting, Mary's finances were more secure, and she wasn't seriously looking to eclipse her discovery of *Squaloraja,* at least not in the same year.

But Mary ended 1830 on a high note, writing to Buckland and De la Beche to announce her unearthing that December of yet another species of plesiosaur, a large-skulled creature with a neck at least three times as long as its head.

> I write to inform you that in the last week I discovered a young Plesiosaurus about half the size of the one the Duke had, it is without exception the most beautiful fossil I have ever seen. The tail and one paddle is wanting (which I hope to get at the first rough sea) every bone in place, in short if it had been made of wax it could not be more beautiful, but...the head is twice as large in proportion as those I have hitherto found. The neck has a most graceful curve and what makes it still more interesting is that resting on the bones of the pelvis is, its Coprolite (fossilized feces) finely illustrated.[11]

Eventually this creature, described by Buckland, was deemed to be a new type of plesiosaur—one with more neck bones than other types. Its scientific name is *Plesiosaurus macrocephalus.*

The skeleton was eventually purchased for £200 by William Willoughby, a Conservative member of Parliament who collected fossil fishes as a hobby. Later Richard Owen, who would become the leading personality in the world of Victorian science, thanked Willoughby for allowing him to examine it and dubbed the specimen the "Hawkins' plesiosaur" after fossil hunter Thomas Hawkins, who had come up with the name for the new creature. Owen failed to mention that it was Mary who originally found the plesiosaur, yet another slight that underscored the continued domination of males in scientific pursuits.[12]

With five major discoveries under her belt, Mary was becoming something of a legend, not just in London but outside Britain as well. Perhaps reflecting her celebrity status, it was around 1830 that Mary decided to change churches.

As mentioned, following Richard Anning's lead, Mary and her family had been faithful Dissenters, regularly attending the Independent Chapel on Coombe Street. No matter what the Sunday, Mary and her mother could be seen walking past the linen draper, glazier, tailor, brewer, stonemason, cabinet maker, and two bakers on their way to the plain brick building with small round windows that had been standing since the 1750s. Despite their meager means, Mary and her mother had always been respectable parishioners and regular contributors to the church's finances. As a child, Mary had found comfort in the words of her pastor, who would have reflected the Dissenters' belief in a sympathetic and benevolent God, one that cherished each person's individuality. Later Mary might have read Dissenters' arguments in favor of the abolition of executions at a time when scarcely anyone questioned the idea of capital punishment.

But by 1830, the Independent Chapel, once among the most thriving and highly regarded Dissenter congregations in Dorset, was close to disbanding, mostly as the result of a dyspeptic pastor, perhaps with little talent for public speaking. Like her father, Mary probably had been drawn to the Dissenters not so much for their religious beliefs as by the fiercely autonomous and questioning nature of the people attending the church. The pastor at the Independent Chapel from 1818 to 1828, John Gleed, had been an inspirational one who supplemented his own salary by collecting and selling fossils. Over the years, Gleed likely had engaged Mary in many a conversation on fossils and how he saw them fitting in with biblical teachings. But when he decided to travel to the United States to participate in the movement against slavery, he was replaced by one Ebenezer Smith, who seems to have pushed at least some parishioners away with his waspish manner.[13]

As a result, Sunday services might have started becoming more of a burden than a comfort. And so, after surely giving the issue great consideration—and perhaps mulling it over with both her mother and brother—Mary decided to shift allegiances to the established Church of England, a move that might have also indicated a hoped-for rise in status and in her place in society.

The decision shouldn't have been that difficult to make. The Church of England had it all in those days: status, wealth, and respectability. Buckland was an Anglican clergyman; so were Conybeare and various other prominent fossil collectors. The Philpot sisters also were Anglican. But throughout the 1700s, some of the country's best-educated and most independent-minded citizens had been Dissenters, courageous people who had stood firm in their refusal to subscribe to the established church teachings.

Mary's religious conversion came during a period of great vigor and change for the Church of England, a time of revivalism when many cathedrals were being remodeled to enable more comfortable and spacious congregational worship. The first part of the nineteenth century were the glory days before the church's role in society was diminished by a series of parliamentary acts allowing nonconformists, Roman Catholics, and other non-Anglicans,.including Dissenters, to vote and to become members of Parliament. State—or civil—registration of births, deaths, and marriages was introduced in the late 1830s, a move that, in theory, removed from the church the important role of charting the progress of people's lives. Yet for many people, one of the Church of England's distinguishing pluses would be its decision to oversee the establishment of 17,000 schools offering education to the poor. Certainly this would have pleased Mary.[14]

With a church register dating back to 1538, Lyme Regis's Anglican church, St. Michael's, could hardly have been more different from the Independent Chapel Mary had grown up in. The Independent Chapel was a low, rambling, modestly decorated space; St. Michael's was tall and airy, with a large expanse of land around it, and with a long history embodied by a wooden pulpit dating from 1613. To accommodate Lyme's increasing population, side galleries had been erected in 1824, which were then filled with box pews. Even then, the local newspaper reported that there was still a lack of seats to accommodate all who wished to attend.

As Mary embarked on a new spiritual journey at a new church, she might have harbored a few doubts about her decision. Although she probably relished attending services at the more prestigious house of worship, in the company of the finest of Lyme Regis's citizens, acerbic Mary still was not about to stop speaking her mind.

9

Spilling Secrets

Anna Maria Pinney moved to Lyme Regis around October 1831, but she almost missed her chance to become friends with Mary Anning. From the very beginning, it seems the privileged teenager's overprotective mother was suspicious of the brusque beachcomber, believing she might somehow sabotage the years of effort sunk into cultivating good morals and habits in the impressionable girl. So wary was she of the renowned fossil hunter that initially she forbade her daughter from going with Mary to the beach unless accompanied by a suitable chaperone. Although Mary was congenial enough, and not impolite to Anna Maria's family, it wasn't going to matter. A frumpish spinster with a sharp tongue and very little education was bound to put off someone like Mrs. Pinney, an upper-class woman of impeccable style.

But Anna Maria saw things differently. From the moment she laid eyes on Mary at her Fossil Depot, it appears she was smitten by the passion and intelligence exhibited by "a woman of such low birth."[1] Undoubtedly Mary was flattered. After showing Anna Maria around the cluttered shop, of which Mary had grown extremely proud, she apparently was so charmed by the teenager's interest in her work that she asked her to go fossil hunting. Beneath the practiced politesse of Anna Maria's small talk, her gray-blue eyes would have blazed with excitement. Of course she wanted to go. The more they talked, the more the restive young lady found herself gravitating toward Mary. If her unusual background wasn't enticement enough, a new friendship would have been a nice distraction from tedious daily lessons under her mother's tutelage. Weeks earlier, Anna Maria had been unsure about her family's move to Lyme Regis. Meeting someone like Mary made the change more palatable.

Anna Maria was the daughter of the entrepreneur J. F. Pinney, who owned sugar plantations in the West Indies and whose wealth was built on the slave trade. He had moved his family from Somerton to a grand house set on a hill in Lyme Regis in 1831. The move was necessary in order to secure his 25-year-old son, William's, campaign to become the first member of Parliament elected to represent Lyme and Charmouth in the forthcoming elections.[2] Although Mrs. Pinney's initial reservations about Mary would never vanish completely, the family ultimately decided that support from the well-known Annings might come in handy. Soon even Mary's brother, Joseph, was out campaigning for William Pinney.

Lyme Regis had long been part of the country's notorious system of "rotten boroughs"—sparsely populated boroughs that could be bought by any politician with a full purse. Indeed, in those days, British Parliaments were elected by only a tiny percentage of the population, part of an unappetizing and undemocratic brew so fraught with corruption there was not even a system in place for registering voters. But the Great Reform Act of 1832, one of the most significant pieces of legislation passed during Mary's lifetime, abolished these boroughs by bringing at least more of the middle classes within the pale of the constitution. During the turbulent 1820s, as a worsening economy swept Britain, so to did a desire to become more democratic. The Reform Act increased the electorate from only 5 percent to 7 percent of the adult male population, but at least it was a step toward widening the franchise. Before the act, the burgeoning populations of thriving new manufacturing hubs had few representatives in Parliament, while barely populated rural areas had far too many.

Despite their wealth, the Pinneys decided William would have a better chance if he ran as a Reform candidate, rather than as a Whig or a Tory, which meant he desperately needed the votes of ordinary people like Joseph Anning and the Annings' middle-class male friends to win. But Anna Maria was not only interested in politics. Anna Maria also enjoyed the fun exchange of gossip, as apparently did Mary. As there are some indications that Mary began finding at least some people of her own rank distasteful after years of being courted by those above her, she probably relished telling tales with her newcomer friend.

Like many young girls of the era, Anna Maria had gotten into the habit of keeping a journal. In it she poured out her heart and soul and, more important, the details of her days with Mary. Much can be gleaned from these journals about Mary's personality and state of mind.

After their first morning of fossil hunting together, Anna Maria recorded that Mary "glories in being afraid of no one, and in saying everything she pleases... she would offend all the world, were she not considered a privileged person... she was very good humored with me, but gossiped and abused almost everyone in Lyme, laughing extremely at the young dandies, saying they were numskulls, not men."

Over the next several months, the two friends spilled all sorts of delicious secrets. Mary filled Anna Maria in on all aspects of her life thus far. She told her of being struck by lightning as a baby. She told Anna Maria of being left, after her father's death, with a mother so distressed she did not attend to her. Mary also told her of her work as a fossil hunter, her scientist friends in the cities, and her rise from obscurity. Together the two shared many lively adventures, some that may have proved to be a bit too lively for the sheltered teenager.

In her journal, she wrote:

Mary goes out just before the waters begin to ebb, and we climbed down places, which I'd have thought impossible to have descended had I been alone. The wind was high, the ground slippery, and the waves beating against the Church Cliffs as we went down. Our dangers were by no means over for when we had clambered to the bottom of the Corporation wall, we had frequently to walk along the Blue Lias cliffs, where there was just room to stand and no more the sea being behind us. In one place we had to make haste to pass between the dashing of two waves, before I knew what she means to do, she caught me with one arm around the waist and carried me for some distance with the same ease as you would a baby.

The journal also indicated that Mary felt she been taken advantage of far too often. Anna Maria noted:

[M]en of learning have sucked her brains, and made a great deal by publishing works, of which she furnished the contents, while she derived none of the advantages. She says she stands still and the world flows by her in a stream, that she likes observing it and discovering the different characters which compose it. But in discovering these characters, she takes most violent likes and dislikes. Associating and being courted by those above her, she frankly owns that the society of her own rank has become distasteful to her, but yet she is very kind and good to all her own relations, and what money she gets by collecting fossils, gives to them or to anyone else that wants it.

One day in November 1831, the two went fossil hunting with another young upper-class woman who was in town, an acquaintance of Anna Maria's named Lucy Oates. Apparently the atmosphere was far from that of an afternoon tea party with the trio fully immersed in fossil hunting. Indeed, that day, unusually, the group talked of nothing else. But Anna Maria wrote in her diary that as soon as Lucy left, Mary turned to her, shook her head, and said: "That young lady has been crossed in love." Anna Maria continued that Mary "must have penetration to have discovered such a thing without any conversation except on geology."

In January 1832, Anna Maria's diary hints that Mary must have confided in her some deep, dark secret. According to Anna Maria,

Mary said she'd been suffering from an illness of the emotional kind for nearly eight years—which coincides with the year of De la Beche's official separation from his wife, Letitia. Although De la Beche had always been a wonderful friend to Mary, and they had much in common despite their differences in social standing, there is no indication that he showed any interest in her romantically, even after his wife's affair and their separation. Yet Mary might have been referring to him when she said that she had been deeply wounded. Anna Maria never recorded the name of the object of Mary's affection, noting merely that Mary had described "a welling up of deep emotion inside her body's core"—a feeling she'd suppressed for many years.

Anna Maria wrote:

[Mary's] wonderful history (which I cannot consider myself at liberty even to write) interested me...I felt the power of the emotions by which she was actuated, and I should have been glad to have possessed sufficient strength of mind to have done the same. An illness of eight years could not bend that spirit, though acute pain supplied the place of health, the bodily anguish was small with what must have been suffered by a proud mind, who had hoped since childhood to see herself removed from her low situation in life, and suddenly saw those hopes blasted by Satanic treachery.

Undoubtedly the typically direct, serious, and hardworking Mary, by then in her early 30s, felt a bit silly divulging her innermost feelings to a sentimental teenager. But there also were plenty of indications that she enjoyed Anna Maria's company and that she felt relieved to share her emotions for a change. Like Mary, Anna Maria was extremely religious, and the pair talked a great deal about topics of a spiritual nature. Anna Maria believed that the idea of creation was too overwhelming for human thought. "To think that life shall never have an end quite fills the mind, but to think of God without a beginning is more than a created being can comprehend," she wrote in her journal. In general, Anna Maria painted a mixed picture of Mary, portraying her as alternately both brave and blunt, often angry, sometimes arrogant, but also willing to help anyone in need.

As if on cue, in mid-1832 Mary unearthed another enormous ichthyosaur, this one a mind-boggling 30 feet long, larger even than a hippopotamus. On the heels of this discovery were many smaller finds. But as she discovered and reconstructed more ichthyosaurs, she also was confronted with more and more buyers who were determined to wheedle their way out of paying what they owed. Over time, the prices of her fossils had started to fall. Twelve years earlier, one of Mary's small skeletons had been auctioned off at Bullock's Museum for £152. By 1832 Mary was offering Adam Sedgwick a specimen she referred to as "the best yet discovered"—a description that could have been hyperbole—for a paltry £35.[3] Soon she was so desperate for cash she found herself dismally satisfied with receiving less than half that for a well-preserved skeleton.

Mary wasn't the only one suffering. By the early 1830s, due to a decrease in the amount he earned from his landholdings, even De la Beche faced actually having to work for a living. Fortunately he was able to make a considerable salary by churning out impressive topographical maps for the Ordinance Survey, beginning with Devon, each of which included pertinent geological information.

Other scientists, too, were making great strides during this period. In 1832 Gideon Mantell uncovered a partial skeleton of a 25-foot armored reptile he would call *Hylaeosaurus* in the Tilgate forests of southern England. Living about 135 million years ago, *Hylaeosaurus* was another nightmarish prehistoric creature, with long spines on its shoulder and rows of armor running down its back. But a year later, in 1833, Mantell moved to Brighton, with an eye towards bolstering his finances, but where ultimately his medical practice suffered so greatly because of the competition for patients that he was nearly destitute. At the same time, in an effort to find proof of an intelligent creator, Buckland busied himself on an important paper that would examine everything from mineral deposits to fossil plants. The paper was one of eight Bridgewater Treatises commissioned by the Reverend Francis Henry the Earl of Bridgewater, a gentleman naturalist, upon his deathbed to explore the power, wisdom, and goodness of God, as manifested in the creation. Buckland worked on the paper for five years beginning in 1831; it was finally published in 1836.

Despite tough times and falling prices, Mary continued to act as a guide for other fossil collectors on tours of the coast, free of charge. Deep down, she probably believed that their visits and her correspondence with them kept her busy and in the limelight.

Coincidentally, one of these fossil collectors, Thomas Hawkins, arrived in Lyme Regis on the exact day in July 1832 that Mary discovered the skull of a giant ichthyosaur. The monstrous giant head was some five feet long, making it the largest ichthyosaur found to date. She uncovered it at the foot of the cliff outside St. Michael's Church, so imprisoned by marl she had to hire some local men to help dig it out. After a careful search, she wasn't able to spot any trace of the rest of the skeleton. And this is where Hawkins came in.[4]

A fossil collector so quirky and mercurial that some later would think him mad. This son of a farmer 11 years younger than Mary had grown up in Glastonbury, in southern England, and was familiar with the same band of Lias that lined the shores of Lyme Regis. Like Richard Anning, Hawkins's father also had kindled an interest in geology in his child, but in this case it was by giving him enough money to purchase his own collection of fossils. By the 1830s, the restless young man had become one of the country's most energetic fossil collectors, building a vast collection with the help of both his drive and his inheritance money.

Hawkins's talents, however, were overshadowed and colored by his tendency toward exaggeration. Forever a showman, Hawkins became well known—and disparaged—for selling a collection of fossils to the British Museum that, without the curators' knowledge, he had tinkered with. So adept was Hawkins at replicating missing parts with plaster that it took a while for anyone to notice. Gideon Mantell was one contemporary who had little regard for such shenanigans. He once described Hawkins as "a very young man who has more money than wit." The questionable techniques of collectors like Hawkins were what had caused Cuvier to question the validity of Mary's plesiosaur so many years earlier.

Mary probably knew quite a bit about Hawkins when he arrived in Lyme Regis. She likely wasn't surprised when this talented collector asked to buy her huge skull—or even when he announced he was going to go after the rest of the skeleton himself. Mary kindly agreed to escort him to the exact spot where she had found the head, all the while assuring him that, in her opinion, she'd scraped out all the bones that there were to be found. But Hawkins was equally convinced the rest of the skeleton was still there somewhere, buried deep within the cliff. He wrote: "Miss Anning had so little faith in my opinion, that she assured me I was at liberty to examine its propriety or otherwise myself."

Hawkins was one of those people who could make even the craziest of ideas sound perfectly plausible. He charmed the landowner into giving him permission to pull down a massive chunk of the cliff and then rustled up an army of men to help with the job. Early the next day, on July 26, before the sun had driven the last of the mist from the hills, Hawkins and his team began humbling the face of Church Cliffs with simple brute force. They toiled hour after hour with a variety of instruments, bringing down one section of the cliff at a time, and by the next day they had attracted a large flock of onlookers.

Hawkins put on quite a show. In all, the men dislodged more than 20,000 loads of earth in the effort to shake loose the skeleton, earth that was later utilized for making a roadway to the beach. Likely much to Mary's chagrin, the men achieved success, exposing the entire skeleton by noon on the third day. Hawkins later described the scene in his memoirs. He said that a hearty cheer rose up from the workmen, which was soon picked up by the spectators, making the surrounding hills and dells ring with noise. Hawkins had been right. He had wanted the skeleton, and he had gotten it. And what a skeleton it was—the largest *Ichthyosaurus communis* found to date. Unfortunately, when the men tried to move the skeleton, embedded in the soft marl, it broke into a myriad of fragments. Even though Hawkins was a rival and had proved her wrong, Mary tried to help as best she could to reunite the pieces, but with limited success. Even so, together they boxed up the fossil fragments—more than 600—as well as the surrounding rock in containers that weighed more than a ton each and shipped them off to Hawkins's home in Glastonbury.

Despite the fact that Hawkins had located the rest of the skeleton, he failed to make much of an impression on Mary, especially in light of his reputation for taking liberties with his fossil finds. She later wrote: "He is such an enthusiast that he makes things as he imagines they ought to be, and not as they are fully found."[5]

Hawkins, however, was effusive in his praise of Mary. In his memoirs, he wrote: "This lady, devoting herself to science, explored the frowning and precipitous cliffs there, when the furious springtide conspired with the howling tempest to overthrow them, and rescued [fossils] from the gaping ocean, sometimes at the peril of her life, the few specimens which originated all the fact and ingenious theories." Apparently Hawkins returned to Lyme Regis the following year, only to excavate another ichthyosaur on the foreshore.

Mary was always keen to have her geological prowess recognized. Even so, she probably couldn't help but feel animosity for someone like Hawkins, a man without financial worries. But he'd publicly praised her, which had been a nice surprise. By then she'd grown accustomed to wealthy men asking for her help, being led by her along the treacherous cliff sides, but without giving her any compensation for her time. Once, after being guided by Mary along the shore, one fossil collector named John Murray wrote how, at the end of their excursion, Mary had handed over all the curiosities they had collected, free of charge. "Thus curious are the simple annals of the poor," he wrote.[6] Even Anna Maria Pinney wrote that Mary complained how "these men of learning have sucked her brains, and made a great deal by publishing works of which she furnished the contents, while she derived none of the advantages."[7] But what was even worse was the careless disregard of museums when handling her discoveries.

Around this time, someone misplaced the coprolites Mary had meticulously preserved along with an ichthyosaur skeleton she had sold to Adam Sedgwick. Not realizing the noteworthiness of the tiny fossilized feces, geologists at Cambridge either lost them or, even worse, absentmindedly threw them away. Later, the British Museum mysteriously misplaced all but the head of Mary's first ichthyosaur, the one that had first made her famous as a young girl, as well as most of the massive one she uncovered in 1832. Mary probably heard about all this later from Buckland, and the careless blunders would have wounded her pride considerably.

At scientific societies, such as the Royal Society and the Geological Society, men still held sway, as women were barred until 1919. Even in the 1830s, men still regarded women as mostly weak and frivolous, more of a hindrance than a help in the scientific arena. To many men, women did not possess the intellectual rigor or perception necessary to engage in serious classroom discussions or rugged fieldwork despite clear evidence to the contrary.

That year, in late 1832, an audacious debate took place in the hallowed halls of the Geological Society, as members wrangled over whether to allow women to attend the popular lectures of Charles Lyell, then professor of geology at King's College. Never before had even a single university in Britain allowed a woman to attend classes or lectures. But Lyell had been asked to present a series of 12 lectures based on his recently published three-volume work, *Principles of Geology,* and men and women across the country were interested. Even Lyell was hesitant to let women attend, though his fiancée, Mary Elizabeth Horner, was an expert on rocks and shells and had contributed to his work greatly during their geological expeditions. Surprisingly, even Buckland harbored his own bias against allowing women to attend scientific meetings and lectures, although he depended heavily on the work of Mary Anning and the assistance of his wife, Mary, who edited and illustrated his books. But eventually Lyell relented; he was forced to change his mind by the unyielding pressure from Roderick Murchison, who threatened not to attend the lectures until allowed to bring along his wife, Charlotte.[8]

Once word got out that women would be allowed they poured into King's College to hear Lyell speak. But after only two lectures the bishop of London stepped in, making it clear that women would remain barred completely from the college. But then, somehow, perhaps at the prodding of his fiancée, Lyell resigned from the faculty, taking his lecture series to the Royal Institution. It was a rare show of dissent in a world that seemed dead set on keeping women out of scientific studies.

Although the Reform Act was making major changes across the country, giving new rights to the middle class, it would still be some time before men would be comfortable with the idea of women becoming geologists. The role of women in a scholarly environment was discussed when the British Association for the Advancement of Science was formed in 1831, with Buckland as president. When the question of whether women should be admitted came up, Buckland wrote a letter to Roderick Murchison saying: "Everyone agrees that, if the meeting is to be of scientific utility, ladies ought not to attend the reading of the papers and especially at Oxford as it would at once turn the thing into a sort of Albemarle-dilettante-meeting, instead of a serious philosophical union of working men."[9] Seemingly even Mary's great friend Buckland still confined women primarily to their traditional roles of wives and mothers.

About 155 miles northeast of Lyme Regis, travel plans were being made in 1832 for one young lady who could not have been more different from Mary. The girl was Princess Victoria, whose one and only visit to Lyme Regis in 1833 (just four years before she would become Queen of England) caused a genuine sensation.

Victoria was born in 1819 into an England mired in economic depression. While most of the country was living hand to mouth, Victoria was spending her formative years in London's Kensington Palace, which not only exuded elegance but all the creature comforts of the early nineteenth century. She was the only child of Edward, Duke of Kent, the fourth son of King George III and Queen Charlotte. At 50 years of age, the duke married Victoria of Saxe-Coburg-Sallfeld. Their only child, Princess Victoria of Kent, was born shortly thereafter. Victoria was less than a year old when her grandfather George III died and her uncle George IV succeeded to the throne. And she was only 11 when he died and her other uncle, the better-liked 64-year-old William IV, took the throne. As the popular William had no legitimate children, Victoria suddenly found herself next in line to take power.

Victoria's father had died when she was just a baby, so mother and daughter lived in near seclusion in a part of Kensington Palace, away from the intrigues of court life. It was a royal arrangement akin to being tucked away from others in the attic of a rich relative's home. Throughout Victoria's childhood, her mother, the Duchess of Kent, was a constant shadow, even sleeping next to the princess at night. But despite being cut off from the outside world, by all accounts Victoria grew up to be an affable girl with a keen intellect.

In 1833 the duchess decided to take advantage of the opportunity afforded by a summer residence at the four-bedroom Norris Castle on Britain's Isle of Wight to show her teenage daughter some of the southern countryside. Not only did the duchess want the future queen to see—and be seen by—her future subjects, she also had wanted to make sure the public bore witness to the conscientious manner in which her daughter was being raised. The duchess, overprotective and not well liked by the rest of the royal family, constantly sought approval for the way in which she was raising her daughter.

On July 8, 1833, the young princess, her mother, her governess, her dog Dash, and her very own bed from Kensington Palace boarded the elegant cutter *Emerald* to make the excursion from Norris Castle to Southampton on the southern coast of England. Accompanying it was the steam packet HMS *Messenger*—the first ever to be anchored at Lyme—which was on hand to tow the cutter when necessary and also to carry the party's carriages and supplies. Only the horses were changed at staging posts.[10]

The next day the royal group moved on by carriage through Dorchester to Melbury for a visit to the grand estate of the third Earl of Ilchester. The earl accompanied the princess and her mother on long walks over rolling hills and patchwork fields. After this visit, the party was scheduled to travel by carriage to Lyme Regis to meet up with the *Emerald*. Although the young princess did not have the opportunity to speak directly with average people, the trip still put the public squarely in front of the young princess's face so that she got at least some sense of the wider world.

Upon arriving in Lyme Regis, the royal procession made an unexplainable error, bypassing Cockmoile Square—where a handful of dignitaries had been waiting for hours to greet the princess—and proceeding directly to the harbor. The festivities were hurriedly relocated, and the mayor, John Hussey, read a welcoming address. After a round of applause, the royal party then passed through a double file of coast guards before embarking from the Cobb's steps onto a floating naval barge that ferried them to the *Emerald*.

Most of those gathered at the Cobb probably had a hard time getting a look at the princess. Likely Mary and her mother and brother were among the hundreds of well-wishers who jammed the shoreline, all straining for a view of the girl who would one day rule their country.[11]

Before they had left London, the duchess had presented her daughter with a small leather-backed notebook to use as a journal during the trip. Victoria dutifully kept a detailed account of all that she witnessed. She wrote of her frustration at not being able to view the scenery very well because of the crush of excited people that fanned out in every direction, everywhere she looked. Lyme, with its distinctive setting, managed to secure at least a brief mention: "a small port but a very pretty one."

Victoria, a symbol of endurance and of the future, was received enthusiastically everywhere she traveled. For Mary, she had looked on a woman preparing for a role of great influence and consequence. Victoria would go on to oversee a vast expansion of the British empire that led to the country becoming a world power. She also would marry Prince Albert, a first cousin born near Coburg, Germany, a man she loved deeply and on whom she relied for advice on every important matter. Ironically, at 34, it was actually Mary who already was breaking new ground at a time when women weren't expected to be literate, let alone capable of scientific discovery.

10

Esteemed Visitors

One bright and beautiful morning in October 1833, Mary was probably feeling in fine form, looking forward to a day on the beach in unusually mild weather. She would have raced down the pathways, Tray clicking along at her heels. She would have anticipated making at least a few decent finds to sell to tourists.

No one was more aware of fossil hunting's dangers than Mary. Time at the seashore had always proven to be an exercise in tide avoidance. She had narrowly cheated disaster on more than one occasion. Her father's own spill on the rocks likely had contributed to his death. Whenever the weather turned sour, as it was prone to do, winds whipping across the Channel could kick up giant walls of rolling white water with the power to pin even the fleetest of foot against rocks and sheer cliff faces. But general words of warning had always fallen flat. Mary was never the type to dwell on the dicey conditions posed by sections of cliffs that appeared ready to buckle at a moment's notice. Always, she swept worries to the back of her mind.

By afternoon, the day had failed to portend anything but complete and utter calm. The beach was probably empty, except for Tray, who never roamed far. Mary had always loved that dog. Not only was he good company during what was on most days a very lonely occupation, but he also was an excellent guard dog, keeping watch over her fossil finds whenever she needed to go for help. Generally, Tray would have spent most of his time sniffing the ground, taking the greatest of pleasures in fishy smells at the water's edge. He probably wasn't much for actually getting into the water, keener to chase the plump, dusky gray seabirds within earshot of Mary.

The atmosphere was as soft and gentle as it could be. But then there it was: a sharp crack. Mary's head would have pitched skyward. A thunderous roar tumbled down the cliff side. She would have known what was coming: a massive torrent of rock and dirt heading straight at her. It was a rapacious bundle, greedily sweeping up anything in its path, a snowball rolling into an epic avalanche. Mary, about to sprint away, saw that the huge slab was aiming right at Tray. Before she knew what was happening, it had swallowed the dog up.

Everything went deathly still as the air thickened with dust. She might have called out Tray's name and listened for his bark, half expecting him to stumble out of the debris. But he didn't. With the air beginning to clear, she would have raced over to where Tray was buried. She knew she didn't have a lot of time as she clawed her way into the mound of earth, praying for a miracle. But there wasn't one. By the time she dug him out, Tray was already dead. She was devastated.

Mary had survived close calls before, but never anything like this. The hungry landslip that had swooped down and swallowed Tray had come dangerously close to killing her. What had triggered such a large collapse of cliff into the sand? Perhaps it was the recent rains. It was anyone's guess. She only knew that she hadn't seen it coming, that it had shot down the cliff without warning, and that her dog had been its victim.

That night Mary would have been inconsolable. Her mother probably would have sympathized but also would have been thankful that Mary had escaped. After all, she might have reasoned, Tray was only a dog. But Mary was grief-stricken.

Many days later, around November 1833, Mary wrote of her grief to Charlotte Murchison, explaining why she hadn't been able to reply more promptly to one of her letters. "Perhaps you will laugh when I say that the death of my old faithful dog has quite upset me, the cliff fell upon him and killed him in a moment before my eyes, and close to my feet...it was but a moment between me and the same fate."[1]

For weeks, Mary would have been miserable. There wasn't much she truly loved in her life, and probably she had loved her dog more than she ever thought she could love anything. It's no surprise that Tray's death left a hole in her heart that stayed with her for many years.

If the landslip hadn't been enough to foul this time in Mary's life, two months later, in early December, she survived another close brush with death. As she headed out to the beach—perhaps one of the first times she'd ventured out after Tray's death—a large, heavy runaway cart came barreling down the steep road, straight toward her. Mary lurched out of the way with mere seconds to spare. Although unhurt, she was pinned against the wall. The cart had come close to crushing her. The incident probably shook her more than she dared reveal to her elderly mother, and left her nerves rattled for the remainder of the day. Elizabeth Philpot described the incident in a letter to Mary Buckland: "Yesterday she had one of her miracle escapes in going to the beach before sunrise and was nearly killed in passing over the bridge by the wheel of a cart which threw her down and crushed her against the wall. Fortunately the cart was stopped in time to allow her being extricated from her most perilous situation."[2]

After weeks of standing by helpless, watching as Mary moped around, Elizabeth Philpot probably would have tried to come up with some sort of grand scheme for boosting her friend's spirits. One way she might have done this is by hosting a Twelfth Night party, a wildly popular tradition in those days designed to mark the end of the Christmas season's festivities. They varied in style and substance but generally combined eating, singing, and drinking, as well as lots of game playing and silly merriment.

The party's focal point was always a special Twelfth Cake, and the host's first order of business was to distribute a slice to every guest as they arrived. Typically such cakes were elaborate, with sugar frosting and gilded paper trimmings, usually densely packed with fruit and decorated with delicate sugar paste figures. Any cake provided by the Philpots would have been no different. But the cake's purpose went beyond looking and tasting good: It was the means by which a king was chosen to preside over the evening's revelry. According to tradition, a bean and a pea were baked into every Twelfth Cake. Whoever got the piece with the bean was crowned king; whoever found the pea was queen. The custom was so well established by this time that anyone put in charge of a celebration's frivolity was called the king of the bean. For the whole of the evening, this bean king reigned supreme. In general, all manner of outrageous behavior was allowed on this one night, the twelfth night after Christmas.

The Philpots would have invited Henry De la Beche—often in Lyme Regis visiting family over the holidays—to any such event, pleasing Mary to no end. Mary's discussions with De la Beche were always lively and enlightening for her; they offered one of the few chances to hear what the gentlemen in London were saying about science. During any encounter, she would have listened to him intently.

At the time, the mid-1830s, De la Beche was more than mildly perturbed by Charles Lyell, who had just dealt a nearly lethal blow to the theory of catastrophism to which De la Beche subscribed. In his dazzling geological opus, *Principles of Geology,* Lyell had speculated on the progressive development of life, arguing as part of uniformitarianism that Earth's history is essentially a directionless one. To Lyell, geological processes were pretty much the same in the present as they were in the past, with landscapes developing over achingly slow and long stretches of time—and without the intermittent meddling of a creator. But others, such as De la Beche and the very religious Buckland, still believed that catastrophic events such as the biblical flood had largely shaped Earth.

Lyell had been going around London, hailing himself as the "spiritual savior of geology," someone with the ability and intelligence to free science "from the old dispensation of Moses."[3] Such claims surely would have irked De la Beche, with his tendencies toward biblical literalism. Likely he would have been frustrated that he wasn't better equipped to compete with Lyell, a trained lawyer, who had always been a more eloquent and persuasive writer and speaker. De la Beche had just finished writing *Researches in Theoretical Geology,* but the book would not make nearly as big a splash as Lyell's thoughtful and provocative study. Published in 1834, De la Beche's work enunciated philosophical treatment of geological questions in an unpretentious style. It would be well received by younger geologists, but not to the same extent as Lyell's work.[4]

Although De la Beche probably would not have admitted it, catastrophism was on its way out by the mid-1830s. Besides Buckland, who himself was having doubts, De la Beche was one of the last few fierce proponents of the theory. As such, he knew he would have to use every weapon at his disposal—including his ability to criticize scientists through cartoons—to fight against Lyell and the uniformitarianism theory.

And Lyell's beliefs made him a prime target. Because Lyell believed in cyclical changes in Earth's climate, he also espoused the possibility that, under the right conditions, some species might make a reappearance. This led Lyell to predict a day when the "huge Iguanadon might reappear in the woods, and the ichthyosaurs in the sea, while pterodactyle might flit again through umbrageous groves of tree-ferns." De la Beche couldn't resist lampooning the controversial expectation in 1830 with a cartoon titled "Awful Changes" in which he depicted an ichthyosaurian professor addressing a toothy class of saurians.[5]

No matter how much De la Beche poked fun at uniformitarianism, though, the theory was important if for no other reason than that it influenced the ideas of Darwin and other important thinkers down the road while also eliminating the catastrophist theory. Indeed, Darwin supported Lyell's idea of uniformitarianism, and it would impact his work—even though Darwin's work never held the same kind of sway over Lyell. Lyell rejected the traditional biblical account of creation, but he also failed to see enough evidence to support the evolutionist's idea that one species of animal could change into another. Although he enjoyed a friendship with Darwin, Lyell would always hold tightly to his belief in a deep division between humans and animals.

If De la Beche didn't have enough on his plate, he also was embroiled in a serious debate with Mary's other good friend Roderick Murchison. Murchison had publicized his finding of a rock formation known as the Greywacke, which he believed predated the appearance of plants. A lot was riding, economically, on Murchison's assertion. At the time, coal was England's most popular power source—and everyone knew coal seams came from ancient swamps. If the Greywacke contained no fossil plants, then it stood to reason that it contained no coal seams. And if this was true, mine owners could save time and money by bypassing these sites and instructing their workers to mine elsewhere. But De la Beche had found proof to the contrary.

While working on his geological maps of Devon for the Ordinance Survey, he found plant fossils in the Carboniferous formation—the formation best known for its coal deposits—that looked very much like materials found in the Greywacke. For De la Beche, much also was at stake. His standing in geological circles depended on a successful outcome to the argument. The two men eventually came to a compromise, deciding that De la Beche's plant fossils could in fact belong to an intermediate system, named the Devonian, lying between the Silurian and the Carboniferous. If nothing else, these types of geological debates helped the world glean a better understanding of Earth's ancient past.

It's not clear what Mary's leanings were when it came to reconciling the fossil evidence with her no doubt staunch belief in the Bible. Certainly the ramifications of her fossil finds would have weighed heavily on the devout churchgoer's mind. When a young boy stopped by her fossil shop one day around this time, apparently he was puzzled as to how fossils could be present so deep inside Earth. He spent some time with Mary, and she was kind enough to help him label his fossils. The boy's father, the Reverend H. W. Rawlins, had no interest in such matters. To the minister, rocks couldn't possibly have been formed gradually over a very long time, because everything had happened according to the Bible. Apparently Mary talked a great deal with the boy, feeding his questions over the possibility of an alternative theory.[6]

For the intimate circle of learned gentlemen geologists, the early nineteenth century was a time of constant bickering, declamations, and displays of temper as each player honed his own opinions. When one alliance was formed, another was broken.

By many accounts, De la Beche enjoyed many friendships but also many debates. He was a solidly handsome, playful man, one without airs, an unmistakable intellectual who could talk about serious subjects without taking himself too seriously. Mary probably liked that about him and would have been riveted by all his accounts of sparring with the great minds of the time.

But after every chance meeting between the two, De la Beche would have immediately returned to London. His work was in the big city, not on the Lyme shoreline, as was Mary's. Every time he left, Mary undoubtedly would have been disappointed.

In the early months of 1834, life returned to normal for Mary. She spent day after day down by the water, slowly but incessantly pacing between the cliffs, searching for more fossils. It was unseasonably cold, with winds so strong on some days they kept whipping the loose strands of hair across her face. Fortunately, at this time another important man breezed into Mary's life, and he turned out to be a dazzling distraction. Most likely the intelligent foreigner was someone from whom Mary could draw enormous energy—as she often did when an educated and ambitious scientist came to town.

Louis Agassiz was a handsome young naturalist from Switzerland, one of Georges Cuvier's earliest protégés. His reason for coming to England in 1834 was an unambiguous one: He wanted to find and study fossil fish, as many of them as he could get his hands on. And one of the best places to find and study fossil fish in all of Britain was in Lyme Regis with Mary Anning and Elizabeth Philpot.

Buckland's fingerprints were all over Agassiz's first journey to England. Not only had Buckland secured financing for the trip and offered him a place to stay, but he had also come up with a daily itinerary for the young naturalist. Agassiz was so moved by both the kindness and the interest shown by the old professor, now 50, that the two became lifelong friends. No doubt Buckland saw a lot of similarities between himself and the zealous Swiss, a born teacher whose lectures were so entertaining they attracted hordes of townspeople who previously had taken little notice of science.

Indeed, the men's personalities, and even their quirks, were a lot alike. Both of them, when they were absorbed in their scientific endeavors, forgot about everything else in the world, even basic hygiene. For example, one morning during Agassiz's visit, the fast friends were so engrossed in a conversation about fossil fish that Buckland forgot to wash out the kettle he used for boiling old bones before brewing their tea in it.[7]

Born in 1807, Agassiz was—like Buckland—the son of a pastor. Growing up in the bucolic Swiss town of Orbe, surrounded by trout-filled lakes in the shadow of the snowy peaks of the Bernese Oberland, Agassiz had spent every waking hour of his leisure time as a youngster collecting, observing, and classifying objects of natural history. He was a proficient fisherman who kept many kinds of fish in a pond he built in his backyard. In time, though, it was the fossils of fish that consumed most of his attention. Over the years his work so impressed Cuvier that the influential figure took the unusual step of sharing his own notes with him. Cuvier died of cholera in 1832, but Agassiz remained a proponent of his work for the rest of his life, never failing to defend his theories of catastrophism whenever they were challenged.[8]

When Agassiz visited Lyme Regis, he couldn't help but be floored at how Elizabeth and Mary had managed to amass 34 new—at least to him—species of fossil fish, remarkably all labeled and identified, and all collected from the environs of Lyme Regis. The women explained how they had carefully matched backbones with teeth found nearby, in the same band of limestone. Amazed by the acumen displayed by these rural women, he thanked them profusely. In his own journal, Agassiz said: "Miss Philpot and Mary Anning have been able to show me with utter certainty which are the ichthyodorulites dorsal fins of sharks that correspond to different types." Agassiz was so grateful that in his pioneering book, *Studies of Fossil Fish,* in which he described an incredible 1,700 species, he thanked Mary and Elizabeth for their assistance in deciphering the secrets of prehistoric fish.[9]

Although his passion was fossil fish, Agassiz soon would be best known as the first person to scientifically propose the idea of an Ice Age, in 1837. For years he had strongly advocated the huge role of glaciers in bringing about physical changes in Earth's crust that had formerly been attributed to the biblical flood.

Eventually, in 1846, Agassiz moved to the United States, where he became a highly respected Harvard professor as well as one of the most important scientists of the nineteenth century. By colorfully describing and drawing prehistoric fish, he provided evidence—probably much to the delight of De la Beche—that life on Earth had survived and progressed despite a series of catastrophes. Through his work, he also helped popularize the study of natural history in America.

Agassiz was one of only a few associates who actually came through for Mary. Indeed, it was Agassiz who, in 1841, would pay Mary her official due by naming a species of fish after her: the *Acrodus anningiae.* In 1844 he named yet another species of fish after her: *Belenostomus anningiae.* Agassiz also named a species of fossil fish after Elizabeth Philpot: *Eugnathus philpotae.* Such acts of respect for women were unheard of among Mary's British colleagues. Every one of her own finds had been named after men. For example, several species were named after Buckland; his giant lizard-like creature was named *Megalosaurus bucklandii*. Many species also were named after Murchison, including *Didymograptus muchisoni, Megalaspidella murchisonae,* and *Murchisonia bilineata*.

Around this time, Mary had another visitor: the infamous Gideon Mantell, the country doctor and famed fossil hunter who had identified the tooth of the Iguanadon more than a decade before. Although a friend of De la Beche's, Mantell was a closer friend of Lyell's, who often clashed with De la Beche. In their letters to one another, Lyell and Mantell revealed a great respect and affection toward one another, with Lyell often expressing concern over Mantell's precarious health.

Mantell might have been a misogynist—and not nearly as amiable as some of the other scientists Mary had encountered. Indeed, his personality seems to have irritated her. Most likely she was familiar with the story of how, in 1822, Mantell's wife, Mary Anne, had found the strange-looking tooth while strolling among the rocks of Sussex, passing time as she waited for her husband to check on one of his patients. When Mary Anne showed the tooth to her husband, he immediately realized that what looked to be a larger version of an iguana tooth was something completely new to science.

Later Mantell wrote a scientific description of the animal using other teeth and bones that had been discovered—without crediting his wife for the initial find. But Mary Anne's tooth had been the key to his discovery of *Iguanodon,* the stock herbivore that grew up to 40 feet long and lived 135 million years ago in the early Cretaceous Period. And this was important in that it was the second dinosaur that was formally named, after *Megalosaurus*. Together with *Megalosaurus,* discovered by Buckland, and *Hylaeosaurus,* discovered by Mantell, the *Iguanodon* was one of the three genera eventually used to define the dinosaur.

According to many accounts, Mantell, trained as an obstetrician, was inspired to study fossils by Mary's discovery of the first ichthyosaur skeleton. It is no surprise, then, that Mantell one day paid a visit to Lyme Regis and in particular to Mary's fossil shop. When he tried to draw Mary out on a variety of geological issues, she apparently held back, refusing to be forthcoming with information even though Mantell was an esteemed member of the Geological Society. Yet she appears to have felt that his questioning was antagonistic. He must have been puzzled by her less-than-enthusiastic reception. But by this time Mary was tired of gentlemen scientists who took credit for other people's work. Mantell left what was perhaps the only unflattering description of Mary at that time:

> We sallied out in quest of Mary Anning, the geological Lioness of the place.... She, the presiding Deity, a prim, pedantic vinegar looking, thin female, shrewd and rather satirical in her conversation. She had no good specimens by her but I purchased a few of the usual Lias fossils.[10]

Actually, Mary had a lot more in common with Mantell than she may have realized. As a shoemaker's son, Mantell was one of the few working-class fossil hunters in a field traditionally dominated by rich collectors. At times, towards the end of his life, and after he moved to Brighton, he was almost rendered indigent. Of course, Mary herself was always struggling, and things would get worse the following year.

In 1835 Mary lost her life savings—more than £300 she had accumulated through the sale of fossils. Various accounts detail what happened. Apparently Mary had invested her money with a man in London after he promised her a major return. But in a devastating twist, he either died suddenly, leaving her with no recourse for getting back her savings, or he ran off with her money.

Again she was left scrambling, frantically trying to sell more and more of her fossils to budget-minded collectors. In July 1835 she wrote to Adam Sedgwick that she had just found a young ichthyosaur—"the smallest I have yet seen, about 1 foot 9 inches in length." But the esteemed professor of geology at Cambridge was himself short of funds at the time. In large part, even for active and successful scientists, the novelty of her fossils was starting to wear off.[11]

Increasingly fretting over her financial status, a kind-hearted Buckland convinced the British Association for the Advancement of Science (BAAS) to provide Mary with an annuity—no minor feat at a time when women were almost never associated with scientific accomplishments. The BAAS raised £200 through private subscriptions, an amount that the British government bolstered with a £300 donation of its own. The annuity of £25 a year—called a Civil List Pension—began in 1838 and was enough to stave off starvation, even if she never again found or sold another fossil.[12] And so, in some ways, it was a banner time for Mary and her mother.

One year gave way to another. Although Mary continued to spend many days on the beach, she failed to make any other great finds. The good news was that she was comfortable enough financially so that the dearth of finds did not lead to despair. But she had labored diligently in the nearly 30 years since she first burst onto the public scene. Her whole life was still devoted to making another big discovery.

More than ever, Mary was confident of her abilities, even taking time out in April 1839 to write to the *Magazine of Natural History* with information about the prehistoric shark *Hybodus* after finding "the most perfect jaw" of the creature earlier that year. She wanted to make sure the magazine was aware that it was she who had discovered, many years before, the existence of fish with both straight and hooked teeth. She also wrote the editor to inform him that a new specimen did not, as indicated by the magazine, belong to a new genus.[13]

Even though she did not make any significant new finds, Mary probably was pleased to learn that her earlier discoveries were continuing to help all the scientists still in the throes of forming their own opinions about Earth's development. Darwin, for example, wrote extensively about ichthyosaurs and plesiosaurs during 1838 and 1839; with both creatures playing a pivotal part in his formation of a theory of evolution. Always, Mary would have been thirsty for knowledge and information. Hearing what others were saying about her finds probably only made her wish to contribute more. By this time, though, Mary was already 40; decades of worry and backbreaking labor had aged her so that she easily could have passed for someone at least 10 years older.

Yet these years were not without their adventures thanks to her extensive experience, which stoked demand for her services. Perhaps the most memorable of these exploits came in September 1839, when the eminent Richard Owen showed up in Lyme for his one and only visit—primarily to see and to court Mary Anning. The era's most celebrated anatomist and paleontologist, it was Owen who would coin the term *dinosaur* in 1842, from the Greek words for "terrible lizard."

Born in Lancaster, England, in 1804, Owen was the son of a merchant who died when he was only five years old. An impressive youngster, Owen took up medicine at the University of Edinburgh before studying anatomy at a private school. A brilliant scientist and naturalist, with a unique talent for interpreting fossils, he eventually was asked to superintend the natural history collection of the British Museum. He loved the job. This priceless, eclectic mishmash of stones, shells, fossils, carvings, minerals, and stuffed mammals gathered by Captain Cook, Sir Hans Sloane, and other explorers and traders was growing—but also falling into disrepair. A distressed Owen embarked on a decades-long campaign to preserve the collection that culminated with the founding of the great Natural History Museum in London in 1881.

Over the years, Owen struggled with a particularly tumultuous and highly publicized relationship with the younger Charles Darwin, one that was initially warm and cordial but rapidly deteriorated because of Owen's fury over Darwin's theories on evolution. A devout Christian all his life, Owen viewed Earth's formation as a series of experiments by the creator, and he had long tried to prove that man and animal were completely separate entities. Imagine his outrage when Darwin sought to highlight a special relationship between man and ape.

A tall, thin, imposing figure with a big forehead and an even bigger ego, Owen was also a charismatic speaker who could hold any audience rapt for hours. He mixed with royalty—even teaching biology to Queen Victoria's children—while also cultivating close relations with the country's most rich and powerful citizens. But Mary doesn't seem to have been intimidated by her visitor's reputation. If anything, she may have resented him. Indeed, she may have been irate that, only a year and a half earlier, Owen had slighted her when he described in great detail to the Geological Society of London her remarkable *Plesiosaurus macrocephalus*—without giving her credit. Still, there was nothing like a visit from a London scientist to lure Mary from the sidelines, and she couldn't help but relish such a diversion.

Owen came to Lyme Regis after attending the 1839 British Association meeting in Bristol and then stopping in on Thomas Hawkins in nearby Somerset. Owen jokingly said in a letter to a friend that he would "take a run down to make love to Mary Anning at Lyme."[14] Apparently many of the learned men also weren't above making a joke or two at the expense of the spirited spinster.

When he arrived in Lyme, Owen was met by Buckland and Conybeare. Owen wrote that "they made me prisoner and drove me off to Axminster, of which Conybeare is the rector."

The next day, Mary gamely agreed to lead Owen, along with Buckland and Conybeare, on a geological excursion along the coast. The sight of these three gentlemen stumbling along the inhospitable cliff faces with Mary—who had the most nimble of feet and knew well the shoreline's many pleasures and perils—surely was an amusing one. The men's clothing was likely clean and formal. Owen even might have donned his regular crisp white shirt with a stiffened collar high above the neck. Mary would have worn her usual heavy but loose clothes and her hardened and battered top hat that looked as if it had been stepped on by a horse.

Owen was impressed by Mary's ability to maneuver up, down, left, right, and even backward over the cliffs. Likely he was embarrassed by how hard he was finding it to keep up. Mary would have enjoyed this chance to show off her agility, stepping lively over the unfriendly terrain, with the educated men straining to follow behind.

Later Owen wrote: "We had a geological excursion with Mary Anning and like to have been swamped by the tide. We were cut off from founding a point, and had to scramble over the cliffs."[15] This episode was one of the few times in his life that Owen actually clambered over cliffs and embankments. He was not a fossil hunter but a comparative anatomist, more used to spending time in a laboratory or library than scrambling about outdoors.

No doubt Owen spent the next morning resting and recovering in bed before visiting the Philpot sisters' collection. He returned to London well before the arrival of the winter of 1839, which, in some ways, was a major misfortune for him. If he had remained in Lyme through December, he would have borne witness to one of the most astounding geological catastrophes of the era.

II

The Earth Moves

or weeks, two families, neighbors who lived just west of Lyme Regis, were growing alarmed by strange goings-on in their neck of town. One night, one family felt their cottage shifting a bit to the left. The next day, after looking around their garden, the other family spotted some fissures, even though everything else appeared pretty much as usual. Then, about a week later, a well-worn pathway seemed to be slightly askew, compared to where it had been before. Finally, a front door refused to close all the way, as it used to.

These families were living in two small cottages set high atop the Undercliff, a raised strip of land a half-mile wide abutting the cliff, covered with dense trees, rare plants, and a jungle of vegetation. Stretching seven and a half miles between Lyme Regis and Seaton, the rough terrain was grazed by cows and sheep, and was overrun with rabbits. The views of the sea were breathtaking, and, on a clear day, the inhabitants were even able to make out the outline of the French coast on the opposite side of the Channel, which is only 21 miles wide at its shortest stretch. This awesome vantage point was essential; the heads of both households made their livings as customs officers who watched for smugglers along the beach and out to sea.

But the shifts in the land surrounding their homes were starting to cast a dark cloud over their bucolic existence. Their growing worries came to a head on Christmas Eve 1839 when one of the families, the Critchards, discovered that parts of the steep and winding path to their house had sunk about a foot. That was all the impetus the two families needed to send them packing—but just for one night. The next morning, Christmas Day, when they stopped by their properties for a look around, all seemed perfectly ordinary. Relieved, they decided to return home for the holiday. William Critchard reassured his nervous wife that, since the cottage had just been built two years ago, its foundation most likely was still simply settling.

By late Christmas night, many of the more affluent families of Lyme Regis had long finished off their geese and flaming plum puddings. William and Mary Buckland and their children probably savored one of the bigger celebrations; their family was staying in town for the holiday season, hoping to do a bit of fossil hunting and visit with local collectors, including Mary, before returning to Oxford. Buckland always liked to get as far away as possible from the university, and his untidy desk, during the winter break. By midnight, most everyone in town was tucked snugly into their beds, stomachs full, including the Critchards and the other custom official's family.

But the night was not to be a quiet one. Indeed, it wasn't long before the peace was shattered by an intense roar, accompanied by a violent movement of the earth. The families' beds swayed back and forth like pendulums, the windows rattled in their frames. Jolted from a deep sleep, the Critchards shot up, eyes wide, and looked around in horror. They couldn't believe it. The floor of their house was rising slowly up toward the ceiling, as if the whole place were closing in on itself. At the same time, the walls were cracking. All the while, noise was almost deafening. At first, they probably thought they were dreaming. Or perhaps they were suffering the ill effects of too much drink. But the Critchards and their neighbors knew enough not to take any chances. They were desperate to get out of there—quickly.

They scrambled out, barely reaching a spot several steps inland before their cottages began slipping down the cliff. They watched as a lifetime of possessions disappeared into the darkness.

When the sun finally came up, the two terrified families finally were able to survey the damage. They had known the coastline was unstable and, certainly, debris had been propelled into the sea before. But never to such a great extent. This time the ground had literally moved underneath them, and they had lost everything.[1]

This landslip was the biggest in recent memory in Britain, perhaps in the country's history. All told, it opened up a chasm three quarters of a mile long, nearly 150 feet deep, and 240 feet wide. By the time it was over, some 8 million tons of rock had split off from the crag and plunged 200 feet into the sea. Boulders the size of cows had broken off and careened toward the water. According to at least one historian, an entire orchard had been transplanted exactly as it had stood, to a spot lower down the cliff side.

That morning—the Boxing Day holiday in England—word spread fast. Soon most of the people of Lyme Regis, including Mary, her brother, and the Bucklands, were on site, creating a carnival atmosphere. People from neighboring hamlets began making their way over, eager to catch a glimpse of the pile of earth that had replaced a precipice high above the sea. Buckland and Mary scoured the scene, searching for fossils as well as any hint of what might have caused the landslip. Buckland would have ascribed the action to months of heavy rain, and he would have been right. Heavy rainfall had saturated the porous chalk cliffs common along this stretch of the shoreline as well as the thick layer of sandstone beneath. The cliffs had been so weakened that they could no longer support the overlying mass of heavy rocks. In recent weeks, the Undercliff on which the two families lived had started to slouch a bit, and then slouch a bit more, before finally crumpling completely in the landslip, opening up a yawning canyon as vast amounts of land stumbled into the sea. So much rock was displaced that the movement threw up a section of the seabed a mile long and some 40 feet high, creating a ridge, or reef, that ran parallel to the shore and enclosed a natural harbor about 25 feet deep. In the years to come, though, this soft sandstone ridge would be completely eroded and washed away.

When Conybeare received word of the landslip, he hastened over from Axminster, joining Buckland and Mary at the scene. The trio spent a great many hours that week going over the catastrophic happenings as they picked through the displaced soil, sand, and chalk, seeking out new treasures. Mary was riveted by the event but lamented the fact that it did not lead to the finding of any unique fossils. She had been sure it would. Around her, though, was a dense phalanx of fossil seekers, scouring the ruins of what had once been a cliff, all looking to her for inspiration, and that counted for something.

The landslip aroused enormous curiosity throughout Lyme Regis and elsewhere. The idea that a landscape could change so dramatically in a heartbeat, right before one's eyes, fired the public's imagination. A few enterprising farmers even began charging to show visitors around the site. Some people visited by paddle steamer, and later a piece of music, the "Landslide Quadrille," was composed to commemorate the event. Even the young Queen Victoria came to survey the spectacle, albeit only from the bay, arriving on her royal yacht from the Isle of Wight. In the end, the landslip was another important reminder of the unpredictable whims of erosion, with cliffs forming over centuries of time or, as in the case of the landslip, in an instant.

After the drama of 1839, the arrival of a new decade probably was something of a quiet letdown. Time passed as it normally did, slowly and monotonously. It wasn't that Mary wouldn't have had anything to keep her busy. In addition to running her fossil shop and caring for her home, she continued her almost daily forays along a foreshore—the most seaward part of the beach—that extended for several miles. She also spent many a day indoors, preparing fossils for shipment to London, Oxford, and other cities. Weeks of careful and precise work were involved in removing the overlying rock from the surface of a skeleton. Mary's main tools continued to be a hammer and chisel. She would have repaired broken bones with animal glue and strengthened weak bones with acacia gum. She would have built wooden frames on which to adequately display and transport the skeletons, possibly by this time using plaster to fill in any gaps. In addition, Mary would have spent time reading dispatches from her many far-off friends. But by now visits from the learned gentlemen were becoming less frequent.

As the new director of London's young and very small Museum of Economic Geology, Henry De la Beche was busy worrying over the lack of space. At the same time, William Conybeare was relishing his new appointment as Bampton lecturer at Oxford, where he was charged with delivering a series of lectures-on Christian theological topics. William Buckland, too, was preoccupied with plans for a geological tour of Scotland. And so, in the fall of 1840, Mary was pleased to be able to offer her assistance to Louis Agassiz by shipping off a few more fossil fish for him to examine, at his request.

Agassiz was in Britain that year to attend a meeting of the British Association for the Advancement of Science, at which he talked up the controversial evidence he'd gathered in support of his idea of an Ice Age. Afterward, a skeptical Roderick Murchison wrote: "Agassiz gave a great field-day on glaciers and I think we shall end in having a compromise between himself and us of the floating icebergs! I spoke against the general application of his history."[2] Mary's fossil fish later would help Agassiz as he mapped out the Rhaetic Bed, a layer of older rock lying beneath the Blue Lias. This was no small contribution; Mary's fish and Agassiz's mapping later helped scientists identify various kinds of ocean creatures that had lived tens of millions of years ago, long before even the dinosaurs.[3]

By this time, Mary was well acquainted with the various modes of transportation available for shipping fossil finds to scientists and buyers. Generally she sent larger specimens by ship from the harbor; the preferred method for shipping smaller items was transport by any willing visitors to Lyme Regis. In one of her last correspondences with Adam Sedgwick, in 1843, Mary said she was sending off the "platydon head" to London by "wagon railroad," the covered wagons used for freight and transportation before the railroad was introduced. Although the London and South Western Railway had been launched in 1838, the first railway did not reach Axminster, some four miles from Lyme Regis, until 1859.

In the early 1840s, Mary was settling into a quiet satisfaction with her life, assured of a steady income and an abiding relationship with God to keep her company. Her deep religious faith would have been great comfort to her when, in early October 1842, her mother, Molly, died from unknown causes, possibly in her sleep. Although she had been in declining health for some time, Molly had been reasonably active until her death at 78.

There are some indications that Molly had always been a bit bemused by her daughter's aspirations, but the two had been constant presences in each other's life, and Molly had shared in Mary's every success and failure. Visitors to Mary's fossil shop later recalled a very old woman wearing a mop cap and a large white apron, somewhat timid, patiently helping customers and gladly showing children around.

Molly once described her daughter as "a history and a mystery"[4]—a strange characterization from the person who knew Mary best. Molly, who had married young, probably never grasped her daughter's desire to be recognized, to be independent, to make a name for herself within a rigidly stratified society in which there was bound to be no place for her. Initially Molly had chastised her husband and her daughter for their foolhardy passion for fossil hunting, considering it a waste of precious time. She accepted Mary's choice of occupations only grudgingly, after realizing it offered the family its best, perhaps only, chance of earning a steady income.

In those first days following her mother's death, Mary might have experienced an exhilarating feeling of freedom. For the first time in her life, she had no one else's needs to consider. But then loneliness surely would have settled in, making itself at home like an unwelcome guest. For anyone, the loss of a parent would have been difficult even in the best of circumstances. But Mary, now 43, had never lived alone, and Molly's death would have made Mary's solitary life even lonelier. Without her mother, she also would have no one to help mind her fossil shop while she was away on the beach. A few days after the funeral, tears might have flowed from Mary's eyes as the enormity of the loss finally hit her. Perhaps her thoughts returned to the memories that hurt her most—memories of her father and all the brothers and sisters she would never know. The loss of both her parents would have made her wonder about the rest of her life and how it would play out. No doubt she lamented the fact that her father hadn't lived to witness her many important discoveries.

Mary wrote a lot throughout the 1840s in what was called a commonplace book—not a journal but, as was popular with women at the time, a notebook filled with quotes and thoughts from her readings. In hers, Mary displayed a distinct fascination not only with fossils and ancient creatures but also with religion and romance and her lot in life. She copied out several pages of information about physics, astronomy, and prehistoric land and sea animals.

Mary also transcribed several verses of poems written by a wide variety of poets. She copied out one, titled "Solitude" by Henry Kirke White, shortly after her mother's death.

> It is that I am all alone...
> Yet in my dreams a form I view
> That thinks on me and loves me, too;
> I start and when the vision's flown,
> I weep that I am all alone.[5]

She also copied a poem called "The Magdalene" by an unknown author that chastised those who looked down on others.

> O turn not such a withering look
> on one who still can feel
> Now by a cold and hard rebuke
> An outcast's missionary zeal![6]

Like many women of this era, Mary's favorite poets included Lord Byron, one of the country's first celebrities—one who was so famous that he received mountains of often-amorous fan mail from anonymous women.

In one of Byron's poems that Mary copied, the poet talks of a man wasted by his past love affairs.

> Tis time this heart should be unmoved
> Since others it hath ceased to move,
> Yet though I cannot be beloved
> Still let me love.
>
> My days are in the yellow leaf
> The flowers and fruits of love are gone—
> The worm, the canker and the grief
> Are mine alone.

Mary also copied an astounding 35 pages of prayers by Thomas Wilson, Bishop of Sodor and Man. The first prayer she noted said: "Afflictions...bring us the nearest way to God." Another said: "Give me a tender compassion for the worries and miseries of my neighbor...that Thou may'st have compassion upon me." She also copied an essay rejecting the way some people used the Bible to oppress women. Some historians have surmised that, at a time when many people thought of religion as a set of rules to be followed to avoid God's wrath, Mary reflected her Dissenter upbringing by viewing God as a compassionate creator who made women out of the same flesh and blood as men and therefore considered the two to be equal.

As Mary passed through these emotionally taxing years, she apparently found great comfort in the musings of others, poring for countless cold, dark evenings over discourses on God, love, and death.

Becoming lost in thought likely was a welcome tonic during a time in which Mary would have been feeling very far removed from all the scientific action in London and Oxford. Over the last few decades, she had become one of the most recognized names in geological circles, working closely with many of Europe's most famous learned gentlemen scientists. With them, she'd debated the meaning of fossils and resolved disagreements, sometimes a bit heatedly. Through it all, she had developed real relationships. But fossils were no longer as novel as they once were, and at least some of her gentlemen friends had gone on to scientific pursuits far removed from geology. She most likely followed the news of their successes as best she could from rural England, but probably not without feeling a tinge of jealousy.

Making a major splash was Richard Owen, the mercurial anatomist, who only three years earlier had pressed Mary for information about fossils. His word *dinosaur* to describe three of the world's terrestrial finds—*Iguanodon, Megalosaurus,* and *Hylaeosaurus*—first appeared in a paper he presented to the British Association for the Advancement of Science in 1842.[7] At around this time he also correctly deduced that the dinosaurs of the prehistoric past had walked upright on two legs rather than on all fours like lizards. After reviewing only the sketchy evidence of three genera, he insightfully concluded that dinosaurs had carried most of their weight with their hind legs. By the early 1840s, Cuvier had been dead for nearly a decade and Owen was being referred to as "Britain's Cuvier,"[8] thanks to his hard work and unrivaled knowledge of anatomy.

Publicity surrounding Owen's 1842 paper led to the first of many outbreaks of what was dubbed by newspapers in Britain as "dinomania." To feed the frenzy, artists were commissioned to put flesh on the bones turning up in Lyme Regis and elsewhere. It was Owen who helped create the first life-size dinosaur reconstructions in collaboration with the renowned sculptor Benjamin Waterhouse Hawkins. Their iron, clay, and stone models of iguanodons, megalosaurs, ichthyosaurs, and other creatures were both beguiling and horrifying to the public.

A God-fearing man, Owen continued to believe that dinosaurs were complex creatures, not primitive ones, and if nothing else they illustrated God's incredible handiwork. In Owen's view, God was the one and only life force, directing the proliferation of cells and tissues and also determining the life span of the individual and of all species—including the dinosaur. Owen saw no evidence for the transformation of one species into another, a process through which, say, a megalosaur might be transformed into a carnivorous mammal.

By this time, Darwin was refining his own controversial theories, telling only the closest of his associates that species came into existence over vast stretches of time through natural selection. Darwin began assembling evidence that this natural selection, played out over numerous generations, results in changes sufficient enough to give rise to a new species. Owen would have none of this. In time he opposed Darwin's assertions and instead shared Buckland's belief that the biological mechanics of all creatures resulted from design and high intelligence. The human brain, he charged, was simply too complex and unique to have evolved from any other creature.

By studying their fossilized remains, Owen also recognized dinosaurs as a distinct new category of creature and laid the framework for their study at a time when only a handful had been discovered. But his major flaw continued to be arrogance, which led him to fail to credit the hard work of other diligent researchers and scientists. For Mary, the real rub would have been that Owen was receiving every accolade in the world while hardly ever stooping to get his own hands dirty, as she had.

Certainly Owen had never been a field man but was one who had learned about anatomy with a scalpel in the confines of a laboratory. It was Mary who, day after day, chiseled away at many of the fossils that were central to Owen's classification of the dinosaur and other creatures. During lectures, Owen often discussed the different types of ichthyosaurs, for example, without ever mentioning that Mary had actually found them. But at the same time he was publicly forthright in his acknowledgment of two men—Buckland and Mantell—and their contributions to his knowledge of anatomy.

Also during the early 1840s, Mary's close friend Henry De la Beche was winning praise in his new role as the first director of the recently established British Geological Survey—the world's first national geological survey. He also finally received approval for a new and improved Museum of Economic Geology, which later became part of the Natural History Museum in London. De la Beche's motives to enlarge the facility were purely pragmatic: Britain required tin, iron, coal, and other raw materials to fire up the engines of the Industrial Revolution, and geologists were perfectly placed to help locate the subterranean deposits of these precious commodities for a needy country. De la Beche was deriving great satisfaction from his personal life as well. His daughter Bessie was marrying businessman Lewis Llewelyn, who would soon become the mayor of Swansea.[9]

At the same time, William Buckland, despite his lingering passion for geology, seemed to be diverting his attention away from Mary's world of rocks and bones to focus on other aspirations. Buckland was on the cusp of an appointment as dean of Westminster, a role that would see him improving the buildings and conditions of Westminster Abbey and its school, particularly the water supplies and sewage, in order to reduce the risk of cholera. With five children surviving to adulthood, Buckland also was consumed with family obligations. He was often described as a kind and affectionate father, one who liked nothing more than having his children milling about him.

Although he was still seeking to make sense of Earth's past, by this time Buckland's support for a worldwide flood was being increasingly challenged. More and more, scientists were viewing the flood as an insignificant regional occurrence, not a global event of any real consequence. But Buckland—along with Conybeare and Murchison and many other scientists—never faltered in their steadfast faith in God and in the Bible, even amid indications that the church was losing its tight grip on science. Although Buckland came to accept Agassiz's ideas about glaciation, he also still believed in the global flood. All that had changed, from his viewpoint, was that he realized the flood may not have been the only phenomenon to put an indelible stamp on the world's rocks, as he had earlier believed. Ironically, though, it was the studies of the devout Richard Owen that debunked the idea of a Noah's Ark in which all animals were gathered up two by two.

Indeed, it was around this time that the greatest of the world's scientists were lavishing praise on Owen for deducing the existence of an unknown gigantic flightless bird in New Zealand on the basis of nothing but the study of a single fragment of bone just six inches long—and from the other side of the world. That bird eventually was called a moa. Owen's studies of the moa made clear the fact that different creatures inhabited different parts of the planet. For example, then and in the past, the animals found in South America were unlike those found in Australia. Scientists began to wonder whether there might be different centers of God's creation. Eventually Owen accepted the idea that there was a progression of life over the years, but he believed the laws behind this progression were divine ones put forth by God at the beginning of time. In other words, God had created not each new species but the laws of design that allowed them to develop and to thrive.

For her part, Mary apparently was forming her own opinions about Earth's development. In a letter to a friend named Dorothea Solly, Mary wrote about the relationship between "Creatures of the former and Present World," displaying a real understanding of the idea that species were evolving and that modern creatures might be descendants of ancient ones.[10]

While many of Mary's learned gentlemen friends were making strides in their fields, Gideon Mantell was mired in misfortune. His failing medical practice was sapping so much of his energy that his geological work suffered. His wife, Mary, left him in 1839, his son sailed off for New Zealand, and his favorite child—a daughter named Hannah—died. There are also some indications that, over the years, Owen had stolen much of Mantell's thunder by both taking credit for his work and also criticizing him to other scientists. Worse, in 1841 Mantell was the victim of a serious carriage accident in London. Somehow he fell from his seat and was dragged a long way across the ground. As a result, he suffered from a debilitating spinal injury and constant pain. But Mantell would always be known for the discovery of *Iguanodon,* the second dinosaur, after *Megalosaurus,* to be named.

On May 11, 1844, back in Lyme Regis, Mary and Elizabeth Philpot might have been discussing at least one or more of these gentlemen and their activities while also lamenting the persistent dearth of new fossil finds. Mary complained that "as we have not had either storms or landslips this last winter there had been but little found." No doubt it had been frustratingly long since her last recorded sale of an unspecified fossil to Adam Sedgwick in 1836 and since her sale of a type of starfish to the British Museum in 1840. Elizabeth, too, might have expressed her frustrations. Indeed, their exchanges were a common routine. But on this day their chatter would have been interrupted. Perhaps someone ran through the streets, pausing only occasionally to knock on doors as he shouted "Fire, fire!" Mary would have looked up and seen black plumes of smoke off in the distance—but not too far off. The sight was sending people into a complete panic, fed by fears that the whole town would go up in flames. Neighbors gathered together their most valuable items before wetting blankets to use to fight off flames.

The fire began with a stray spark from an open hearth at the George Inn on the eastern side of Coombe Street, which runs north and south through the town. Fanned by a northeasterly wind, the fire caught on and spread, quickly consuming much of the eastern side of the street. Eventually it leapt up the street, burning buildings to the west before crossing the river and proceeding north, where it destroyed the medieval Customs House and the luxurious Three Cups Inn, the hotel where Buckland and so many of Mary's customers had stayed over the years. There the fire ripped through the thatch of the stables, which burned furiously. While the horses were being rescued, the flames spread to the inn. To Mary and Elizabeth, it must have seemed as if the fire raged on for hours.

At that time in rural towns such as Lyme Regis, firefighting consisted largely of "bucket brigades"—able-bodied men lining up and dumping pails of water on a blaze. The first steam-powered fire trucks had been introduced in London in the late 1820s, but it would be decades before they reached rural England. During the fire, several bucket brigades worked tirelessly, hefting buckets—each containing about three gallons of well water—to one blaze after another. Empty buckets were returned by another line of boys and women to be refilled. But it wasn't water that finally stopped the fire. Eventually the firefighters pulled down and tore apart the town's meat market, called the shambles, in what would be a successful attempt to cut off the blaze. The creation of a firebreak by means of demolition, in order to deprive the fire of fuel, was and still is a widely used firefighting technique.

By the time the fire had burned itself out, much of the old town had been destroyed, including many of the buildings close to Cockmoile Square, where Mary had been born. Fortunately, the Assembly Rooms were spared, as was Aveline House, owned by De la Beche's family, as well as the Philpots' home.

There would be other fires in Lyme Regis. Towns across England were like tinderboxes in those days—all just infernos waiting to happen. Thatched roofs, narrow streets, open fires, and the use of candles for lighting were a recipe for disaster.

Mary and her home had escaped the flames, but the fire forever changed the character of the town she knew so well. She was saddened by the destruction of much that was so comfortingly familiar to her. "I do regret the Old Clock that had stood for Centuries," she wrote a friend.[11]

Despite the fire, the year 1844 did have a high note. Mary would have been in good spirits—having just sold an ichthyosaur and a plesiosaur to an American geologist, Thomas Wilson, who then donated it to the Academy of Natural Sciences of Philadelphia—when, on July 1, King Frederick Augustus of Saxony—now part of Germany—visited Lyme. As usual, he was accompanied by his personal physician, the highly respected Carl Gustav Carus. Carus was a professional of many talents—a doctor, scientist, and naturalist—who is best known for the concept of the vertebrate archetype. This idea was key to Darwin's development of the theory of evolution, providing a direct steppingstone to the notion of evolutionary ancestors.

In his journal of the trip, Carus described a visit to Mary's fossil shop: "We had alighted from the carriage, and were proceeding along on foot, when we fell in with a shop in which the most remarkable petrifactions and fossil remains—the head of an ichthyosaurus, beautiful ammonites, etc. were exhibited in the window. We entered and found a little shop and adjoining chamber completely filled with fossil productions of the coast."[12]

He also wrote a description of how the fossils were found in the area: "It is a piece of great fortune for the collectors when the heavy winter rains loosen and bring down masses of the projecting coast. When such a fall takes place, the most splendid and rarest fossils are brought to light, and made accessible almost without labor on their part. In the course of the past winter there had been no very favorable slips; the stock of fossils on hand was therefore smaller than usual."

Carus and the king searched the cluttered curiosity shop, keen to bring back something of substance to Saxony. Finally, Carus wrote in his journal that the king "found in the shop a large slab of blackish clay, in which a perfect ichthyosaurus of at least six feet was embedded. The specimen would have been a great acquisition for many of the cabinets of Natural History on the Continent, and I consider the price demanded—£15—as very moderate." The king purchased the specimen for his natural history collection housed in Dresden.

The visit obviously made quite an impact on the party. Carus later wrote that, in the end, he was "anxious at all events to write down the address, and the woman who kept the shop, for it was a woman who had devoted herself to this scientific pursuit [and] with a firm hand wrote her name, Mary Anning, in my pocketbook."

When she handed the book back to Carus, she smiled confidently. He looked down at what this lower-class woman in plain clothes had written beside her name: "I am well known throughout the whole of Europe." Perhaps he raised his eyebrows at such a bold declaration. Mary turned and walked away.

12

The Making of a Legend

$\mathcal{E}$ very time Mary heard the waves gently caressing the shore, she was in her element. Even more so every time the first big gusts of a storm blew in, sending waves crashing across the shoreline's pebbles in a fury. The onslaught of a violent storm still moved her into a happy tizzy as she anxiously awaited the next low tide. As always, her pulse raced and her muscles grew tense. Like her father, Mary would still race from the house with one thought pressed into her mind: new fossils. Even after so many years, the worries of the world fell away, narrowing into a solitary single goal, whenever she reached the beach after a batch of wicked weather.

It was early 1845, just before Mary's forty-sixth birthday, a time when a man in her position might have been resting on his laurels. By now Mary had earned the respect of most of her gentlemen peers in the scientific arena. Some of the most famous people in Europe clamored to meet her, to exchange ideas with her, and to see her finds for themselves. Mary's steady stream of discoveries, begun when she was 12, had laid the foundations for groundbreaking reports on a broad array of bizarre prehistoric creatures. But perhaps most important of all, at least to Mary, was that Richard Anning would have been beyond proud of the way in which his cherished daughter had stepped up to the plate in his absence, caring for Molly and creating a real and lasting stability for the entire family, including Joseph, his wife, and their children.

But Mary wasn't reveling in any of this. The truth was, according to most accounts of this period, she wasn't feeling very well. More and more, she was lethargic, drained of energy. Probably she also was pale and losing weight. Most worrisome, though, might have been a real and nagging pain in her breast centered around a large lump. Every day she might have run her fingers over it to see if it was still there. And it always was. She might have told Elizabeth Philpot about it, and most likely her friend would have encouraged Mary to see a doctor.

In those days, a diagnosis of breast cancer was a death sentence. Indeed, the diagnosis carried such a dismal prognosis that most doctors refused to inform the patient, preferring to deliver the bad news to family members instead. It would have been no different for Mary, who likely was dismayed at the lack of information and treatment options. At the time, no one was sure what caused breast cancer. Some thought sour milk or breastfeeding might be to blame; others thought it was more likely related to one's sex life—or lack thereof. Until the nineteenth century, the only remedy on offer was extremely crude and risky surgery, before which doctors rarely even washed their hands. It was not until the end of the 1800s that mastectomies finally started to become common procedures. But during Mary's time, women diagnosed with breast cancer often were considered to be such lost causes they often were even refused extensive or prolonged care or admission to hospitals.

Throughout 1846, Mary grew weaker and sicker, her hair often running with sweat, her breast throbbing. Even so, she probably still felt drawn to her beloved beach, wavering on whether she had the strength to return. Most likely, concerned friends would have urged her to rest. Those closest to her would have seen that she was in no shape to chase down fossils, especially in inclement weather. While Mary was forced to spend more and more time in her bed, the town continued to prosper from the tourist trade she had helped to build. Outside of Lyme Regis, the repercussions of the Industrial Revolution were many. The introduction of steam power drew an ever-mounting number of people from the countryside into rapidly expanding cities. But with the availability of work in Britain's mills and factories, the public was growing increasingly concerned about poor working conditions. Several labor measures were passed, including the Ten Hour Act of 1847 that limited the number of hours a woman or child could work in any given day.

Undoubtedly Mary was focused mostly on her own insular world, marred by her pain, which grew steadily worse with each passing month. Apparently her only relief was laudanum, an extremely popular drug during the Victorian period in both Europe and the United States. The widely used painkiller—a mixture of alcohol and opium derivatives—was prescribed for just about everything, from fevers, to headaches, to tuberculosis, to menstrual cramps. Some fashionably minded women even used it to derive the pallid complexion associated with tuberculosis, as both frailty and paleness were in vogue during this time. Others used it to alleviate the discomfort of their tight-laced corsets. Cheaper than gin, it was not uncommon during the 1800s for working-class men and women laborers to binge on laudanum to celebrate the end of a hard week of work. Members of the middle class commonly used laudanum as well, and even many upper-class men and women developed a dependency on the drug. Laudanum use even insinuated itself into the lives of many famous and highly respected writers and authors including Charles Dickens, Louisa May Alcott, and Elizabeth Barrett Browning.

But laudanum has many side effects, including dizziness, sleepiness, depression, confusion, restlessness, and clammy hands. According to some historical accounts, at least some people in Lyme Regis thought Mary had started drinking during this time, unaware of her reliance on laudanum.[1]

Amazingly, Mary apparently carried on working, at least on some of her better days, and remained remarkably alert, despite her suffering. Sometime during 1846 or perhaps even early 1847, she copied out—or wrote herself—two lengthy, comical poems. One, called "The Complaint of a Sunbeam against Dr. Faraday," looked at the work of British pharmacist Michael Faraday—who discovered electromagnetic induction—through the eyes of an abused sunbeam. The other poem was a limerick celebrating the geological accomplishments of her old friend Roderick Murchison and in particular the knighthood bestowed on him in 1846 by the British prime minister, Robert Peel. This poem also poked fun at Adam Sedgwick, the Anglican clergyman and Buckland's counterpart at Cambridge, for holding tight to old theories of Earth's formation despite the evidence found by Mary and others, as well as at her good friends Buckland and Agassiz. Here are just a few of the poem's verses:

> Let Sedgwick say how things began
> Defend the old Creation plan
> And smash the new one,—if he can.
>   Sir Roderick.
>
> Let Buckland set the land to rights
> Find meat in peas, and starch in blights,
> And future food in coprolites.
>   Sir Roderick.
>
> Let Agassiz appreciate tails
> And like the Virgin hold the scales
> And Owen draw the teeth of whales
>   Sir Roderick.
>
> Take then they orders hard to spell
> And titles more than man can spell
> I wish all such were earn't so well
>   Sir Roderick.[2]

In July 1846, Mary was paid some due, at least locally, when she was named the first honorary member of the new Dorset County Museum in Dorchester, established the same year. After Buckland received word of her illness, he persuaded the members of the Geological Society in London to create a special new fund for her. During this time, although he was now serving as dean of Westminster, Buckland still held great sway over members of the society. But Mary needed more than money. In the face of inadequate treatment, Mary Anning finally succumbed to breast cancer, dying on Tuesday, March 9, 1847, after having endured serious pain for at least two years.[3]

As De la Beche told members of the Geological Society after her death, she bore with fortitude the progress of her cancer. Little has been written about Mary's death. Perhaps she died alone. It's not clear how close Mary remained to her brother, Joseph, who died two years later, in 1849, at the age of 54. Like Mary, he had switched from the Dissenter Church, serving as church warden at St. Michael's Church from 1844 to 1846. Three of his infant children share his grave in the churchyard, near Mary's plot. It is known that Mary had very few close friends at the time of her death. Anna Maria Pinney had never married and had died when she was fairly young. Elizabeth Philpot died a decade after Mary, her obituary reminding the world that she had gone "upon the Lias shore with Mary Anning almost daily."

The Reverend Frederick Parry Hodges, who had been the vicar at St. Michael's since 1833, conducted Mary's funeral instead of passing the job off to a curate, as he might normally have done. Her body was buried in the yard outside St. Michael's that overlooks the sea, at the top of the eroding Church Cliffs she had combed so often. Members of the church and the Geological Society in London paid tribute to Mary with a stained-glass window at St. Michael's that portrays six acts of mercy from the Bible: visiting the sick, feeding the hungry, giving drink to the thirsty, clothing to the naked, housing the homeless, and visiting orphans. The window was dedicated "in commemoration of her usefulness in furthering the science of geology...her benevolence of heart, and integrity of life." It was a lovely way to commemorate the humble woman who started out her life as a Dissenter but ended it as an Anglican.

Not long after her death, glowing accolades poured in, and Mary's obituary was published by the *Quarterly Journal of the Geological Society*. Some five decades later, writer Terry Sullivan was inspired by Mary's life story to compose the popular tongue-twister:

> She sells seashells on the seashore
> The shells she sells are seashells, I'm sure
> So if she sells seashells on the seashore
> Then I'm sure she sells seashore shells.[4]

Even London's literary giant, Charles Dickens—who had written about the despair Mary had seen for herself in the city with an insight born of his own experiences as a child—knew of her life. He wrote his own impressions of Mary—and of those around her—in his weekly literary magazine *All the Year Round*. In it he praised her "good stubborn English perseverance," her intuition, her courage, physical and mental, in the face of locals who initially mocked her eccentricity. He concluded his overview of her life with "the carpenter's daughter has won a name for herself and has deserved to win it."

But most important, in his article Dickens highlighted the strange lack of appreciation and the overall disregard for Mary from those in her own town.

In her own neighborhood, Miss Anning was far from being a prophetess. Those who had derided her when she began her researches, now turned and laughed at her as an uneducated assuming person, who had made one good chance hit. Dr. Buckland and Professor Owen and others knew her worth, and valued her accordingly; but she met with little sympathy in her own town, and the highest tribute which that magniloquent guidebook "The Beauties of Lyme Regis" can offer her is to assure us that "her death was, in a pecuniary point, a great loss to the place, as her presence attracted a large number of distinguished visitors."

According to Dickens, "quick returns are the thing at Lyme."[5]

Although some, like the Lyme Regis historian George Roberts, recognized her uniqueness, even calling Mary "a Helen {from Greek mythology} to the geologists," "the progressive discovery of the structure of the Ichthyosaurus taking about the same number of years as the siege of Troy,"[6] others in town apparently still looked down on her humble background, blunt tongue, and woeful—often manly—appearance.

A year after Mary left this world and the smattering of tributes began fading away, a distinguished-looking man took to the podium before the Geological Society in London, readying himself for his presidential address. The man was Henry De la Beche. As was traditional, De la Beche—elected as president of the society in both 1847 and 1848—used part of his address to mark the lives of those Fellows who had died during the previous year.

He must have ticked off a litany of agenda items before coming out with a most unusual and touching accolade—one that marked the first time such an honor had ever been bestowed upon a woman in this kind of forum.

> I cannot close this notice of our losses by death without advertising to that of one, who though not placed among even the easier classes of society, but one who had to earn her daily bread by her labor, yet contributed by her talents and untiring researches in no small degree to our knowledge of the great Enalio-Saurians, and other forms of organic life entombed in the vicinity of Lyme Regis...there are those among us in this room who know well how to appreciate the skill she employed (from her knowledge of the various works as they appeared on the subject) in developing the remains of the many fine skeletons of Ichthyosauri and Plesiosauri, which without her care would never have been presented to

the comparative anatomists in the uninjured form so desirable for their examination.

The talents and good conduct of Mary Anning made her many friends; she received a small sum of money for her services, at the intercession of a member of this Society with Lord Melbourne, when that nobleman was premier. This, with some additional aid, was expended upon an annuity, and with it, the kind assistance of friends at Lyme Regis, and some little aid derived from the sale of fossils, when her health permitted, she bore with fortitude the progress of a cancer on her breast, until she finally sunk beneath its ravages on the 9th of March 1847.[7]

It was a remarkable and loving eulogy, delivered by a kind friend who had always been enormously sympathetic to Mary's plight and circumstances. The eulogy also was amazing considering that women were not accepted as Fellows of the Geological Society until more than 50 years later, in 1904. No doubt, Mary would have been both delighted and gratified.

By this time, De la Beche had published numerous memoirs on English geology and was so highly regarded that he was knighted later in the year by Queen Victoria in recognition of his geological achievements. After losing the nice flow of income from his landholdings in Jamaica, he had helped turn geology from a gentleman's occupation into more of a government service. Indeed, De la Beche's constant lobbying led to the government providing employment for geologists in areas outside the traditional academic circles, with many moving to India, Canada, and Australia. Again and again, De la Beche urged the government to fund scientific activities, casting it as essential for boosting the nation's standing in the world. Near the end of his life De la Beche was awarded the Wollaston Medal, the highest honor handed out by the Geological Society. After decades of fieldwork and beachcombing, he suffered from partial paralysis in his last year, but managed to work until a few days before his death in 1855.

During De la Beche's eulogy of Mary, he mentioned a member who had persuaded Lord Melbourne to come to Mary's assistance; that member was none other than his good friend William Buckland.

In the years after Mary's death, Buckland focused his attention nearly exclusively on Westminster Abbey and Westminster School. Unfortunately, around 1850, he suffered increasingly from mental illness. In 1856 he ended his days in a mental asylum, apparently dying from tuberculosis, which had infected his brain. Mary Buckland followed her husband to the grave a year later. Throughout Mary Anning's lifetime, Buckland had been one of her most ardent supporters. Toward the end of his life, while he was in the asylum, he fondly recalled the days when Mary, only a teenager, had guided him along the treacherous cliff sides of Lyme Regis.

Buckland was never able to relinquish his belief in a world carefully built under the influence of a divine creator, even as Darwin was shoring up his evolutionary theories with more and more evidence. Although Darwin's *Origin of Species* was not published until 1859, he had long held the opinion that man was not specially created by God but instead may have evolved from apes. This view tormented Darwin so greatly that he was afraid during the 1840s and early 1850s to confess his findings to anyone but his closest and most trusted friends. Both Buckland and his one-time student Roderick Murchison continued for many years to fulminate in favor of the catastrophist position, which believed that events of the past are directly guided by God and have no counterparts in the present world.

Although remaining a highly respected geologist throughout his life, Murchison, who was knighted in 1846, never fully recovered from the death in 1869 of his beautiful and charming wife, Charlotte, who had always remained Mary's faithful friend. Before his death in 1871, he founded a chair of geology and mineralogy at the University of Edinburgh.

Another of Buckland's former students, Charles Lyell, continued his commitment to the nondirectional view of Earth known as uniformitarianism. But he always insisted—in opposition to Darwin—that there was no evidence in the fossil record of a progression from simple organisms to more complex ones. Knighted in 1848, Lyell died in 1875 and was buried at Westminster Abbey.

Another keen geologist during Mary's lifetime, William Conybeare, abandoned geology altogether in order to funnel all of his energies into matters of religion. The vicar at Axminster died 10 years after Mary. At the same time, another of Mary's avid fans, Louis Agassiz, traveled to the United States, remaining there until he himself died in Massachusetts in 1873.

Gideon Mantell, one of the rare gentlemen geologists who failed to make much of an impression on Mary, died a defeated man in 1852 after accidentally overdosing on pain medicine. During his lifetime, he had amassed a considerable amount of fossil evidence and even in death continued to contribute to the advancement of science. In an ironic twist, a warped part of Mantell's spine went on display as part of Richard Owen's collection at the Royal College of Surgeons, where it remained until it was destroyed by a German bomb during the blitz of World War II.

Although initially friends, Owen ultimately turned out to be Mantell's nemesis, one who often challenged Mantell's assertions and even tried to discredit him on several occasions.

In the years after Mary's death, it was Owen whose star would shine most bright. He used his growing cachet to agitate even more for a dedicated museum that would separate "the works of God from the works of Man."[8] Owen, whose career had yet to peak in 1847, believed that this new museum should be the best in the world and that it should display as much of the natural history collection stored at the British Museum as possible because "the British public has a right to see comprehensive displays of all the species making up the natural world." Others disagreed. Thomas Huxley, secretary of the Royal Society, wondered what the public could possibly gain from being allowed to inspect every single species of beetle that was known to mankind, and wanted a much smaller museum.

But Owen got his way and, on Easter Monday 1881—34 years after Mary's death—the doors of London's new Natural History

Museum on Cromwell Road were finally thrown open, attracting a crush of 17,500 visitors. Designed by the famous English architect Alfred Waterhouse, the museum was one of the finest examples of Victorian exuberance, with its grand facade covered in terra-cotta tiles, and topped with turreted spires, so that it looks something like a church. The *Times* of London dubbed it "a true Temple of Nature, showing, as it should, the Beauty of Holiness."⁹ Although Owen reveled in this lasting monument to his scientific ambitions, he died in 1892 a broken and lonely old man, after his only son committed suicide in 1886 and following the death of his wife. But no doubt his grand museum—with its mile of wall space and four acres of flooring—was considered a remarkable achievement, educating one and all.

Even now the magnificent museum captivates large audiences. Today it draws upward of 3 million visitors a year, a sign of the public's continued fascination with natural history. A writer named H. V. Morton once described the museum as a "gothic building that gives the visitor...the impression that the zoo has escaped from Regent's Park and taken refuge in a cathedral."¹⁰ The description is right on. The labyrinthine galleries themselves are home to some 70 million species, from immense pickled squids and mammoth skeletons to tiny lichens and seed plants. And the museum is still being updated and expanded. The opening of the new Darwin Center in September 2009 allows visitors to explore another eight floors of new plant and insect specimens.

On most days, a long line still forms at the front door as visitors wait for the museum to open at 10 A.M. Although many are here to see the museum's famous dinosaur skeletons, including an 85-foot-long *Diplodocus* that greets visitors as they walk through the central hall, there are all sorts of other fascinating items to look at, from tiny live leaf-cutter ants to specimens of large mammals and the now extinct Mauritius dodo.

Walk down one corridor and the stony gaze of the skull of an ichthyosaur—Mary's first big discovery—stares at you. Walk down another and find an oil painting of Mary by an unknown artist. Although hardly flattering, it's one of the only representations of Mary Anning available today. Mary's first *Plesiosaurus,* unearthed in 1823, also is mounted in one of the galleries, near the museum's main restaurant and the ever-popular Creepy Crawlies permanent exhibition. School groups often visit here; most likely few of the children are aware of the history that lurks nearby.

Lyme Regis is a three- to four-hour drive from London, past the perfect prehistoric geometry of Stonehenge. It remains very much like the village in which Mary lived her entire life. With a population of 3,600, the quaint town is still a haven for both fossil seekers and tourists, with tearooms, craft shops, and antique stores that tumble along cozy narrow streets behind Georgian facades to a promenade hemmed in by thatched beach cottages and packed fish-and-chips restaurants. The churches and many other buildings Mary knew are still standing. Although her fossil shop has been demolished, a plethora of other stores selling fossils and rocks, as well as workshops laden with gray dust, can be perused along many of the streets crisscrossing the town.

Out to sea, colorful fishing boats ply the waters. The rounded harbor wall of the Cobb—on which the town still depends for protection—to this day inspires artists and looks much as it did after being rebuilt during Mary's lifetime.

At the small Lyme Regis Museum, built on the site of the home Mary grew up in, a permanent exhibition tells her story. And although her thatched house on Broad Street is no longer there, a plaque designates the location.

The town's coastline that was so combed by Mary is today part of a 95-mile stretch of dramatic shoreline, called the Jurassic Coast, that enjoys the distinction of being England's first natural UNESCO World Heritage Site. The award, granted in 2001, puts the coastline in the same league as the Grand Canyon and the Great Barrier Reef. The UNESCO declaration of December 13, 2001 called it "an outstanding example representing major stages of the earth's history, including the record of life, significant ongoing geological processes in the development of landforms, and significant geomorphic or physiographic features."

It is along this coastline—fanning out both to the east and west of Lyme Regis—that Mary's spirit truly lives on, watching over the Wellington boot-clad fossil hunters, including many young girls, who continue to roam this most inhospitable yet lovely of environments for tiny trilobites, ammonites, belemnites, and, if they're lucky, the remnants of some kind of never-before-seen monster.

Epilogue

> We know not the millionth part of the wonders of this
> beautiful world.
>
> —Gideon Mantell, 1822

On a bright summer's day in 1999, the streets of Mary's youth were overrun with anxious moms and dads, trying to herd their overexcited children to the beach while weighed down with towels, beach chairs, and inflatable toys. Little did many of them know, but a full program of events was under way at the nearby Lyme Regis Philpot Museum. Billed as a "celebration in honour of the first woman geologist," one designed to coincide with the bicentennial of her birth, the day of activities had been organized to honor the life of one Mary Anning. In the morning, an actress portrayed the renowned fossil hunter while, throughout the afternoon, present-day collectors gave talks about their own recent finds. At one point, a basket maker came out to demonstrate how to craft the sturdy containers Mary once used to hold her finds. At a crafts stall, children could be kept busy creating their own artwork. The official convener of the symposium was Sir Crispin Tickell, patron of the museum, former British permanent representative to the United Nations, expert on climate change, and—perhaps most notably for the day's purposes—a great-great-great-nephew of Mary Anning. To most in attendance, the highlights of the day were a keynote speech titled "Mary Anning's Life and Times" followed by an afternoon tea in the back garden of the home of John Fowles, a longtime Mary Anning aficionado whose 1969 novel *The French Lieutenant's Woman* was set in Lyme Regis. Meryl Streep portrayed the unhappy Victorian lover in the 1981 film of the same title and was shown standing at the end of the Cobb. Fowles died in his home in Lyme Regis in 2005.

Ten years after the celebration, in 2009, visitors to Lyme Regis will find a permanent exhibition on Mary Anning laid out on the second floor of the small museum, which ironically stands on the site of Mary's birth. As part of the exhibition, there is a reproduction of Henry De la Beche's painting of Mary's discoveries, the one so instrumental in bringing her family some much-needed income in the 1830s. There's also a fun piece of furniture known as Buckland's Dinosaur Poo Table, made from a slab of inlaid coprolites. From time to time, the museum also hosts various temporary exhibitions on topics having nothing to do with Mary.

As for Mary's actual discoveries, they aren't there. Some are housed in various institutions across the country, but too many have been lost or misplaced. The skull of the first ichthyosaur found by Mary's brother, Joseph, in 1811 is on display at the Natural History Museum in London. The rest of the 17-foot skeleton is nowhere to be found. Although the British Museum purchased the whole specimen from William Bullock in 1819, it either neglected to keep the body or else somehow lost track of it over the years. Mary's first *Plesiosaurus giganteus,* discovered in December 1823, is also at the Natural History Museum, but her *Squaloraja polyspondyla* was destroyed in World War II. At least a few bones of Mary's *Pterodactylus macronyx,* found in 1828, are kept at the Natural History Museum, but they are not on display, while her *Plesiosaurus macrocephalus,* found in 1830, is also at the Natural History Museum. Mary's commonplace book, known as the Fourth Notebook, survives, but unfortunately her first, second, and third notebooks do not.

All in all, despite the efforts of the Lyme Regis Philpot Museum and of a handful of seemingly tireless historians and academics, there has been scant recognition of Mary's contributions throughout the world, and her name is not widely known, even in her home country. Ask any Briton on the street if they've heard of Mary Anning and, more often than not, the answer is no.

But it was Mary who, after spending day after day chiseling curiosities out of the chalk cliffs of southern England in all sorts of weather, laid the groundwork for the theory of evolution, not to mention nearly two centuries of spectacular discoveries in the still-evolving worlds of paleontology and geology. One can only imagine what Mary would think about the fate of her finds as well as about all that has been discovered since her death in 1847. No doubt she would have been overwhelmed by all that has transpired since her days spent skimming through the dark marl, donned in long skirts and a top hat.

When it comes to prehistoric creatures, none has parlayed its fame longer into the afterlife than dinosaurs, which have starred in more than 150 films worldwide. But despite many waves of dinomania, it was fossils like the ones Mary discovered that scientists relied on the most in helping them to decipher the global geologic record. Darwin, for example, was inspired by fossils, not dinosaurs. Indeed, the word *dinosaur* appears nowhere in his *Origin of Species*.

And, like the dinosaurs, many of Mary's creatures have gone on to achieve great fame in these last two centuries. Her long-necked plesiosaur—the one that perplexed even the great Georges Cuvier—has appeared in a number of children's books and several films, including in Jules Verne's novel *Journey to the Center of the Earth*. Modern legends of aquatic monsters, such as the Loch Ness monster, also are sometimes explained as sightings of extant populations of plesiosaurs. In 1977 the discovery of a mysterious two-ton corpse with flippers and with what appeared to be a long neck by the Japanese fishing trawler *Zuiyo Maru* off the coast of New Zealand sparked a plesiosaur craze in Japan. Fishermen joked that the 32-foot whalelike creature might be a monster; others seriously suggested it was another type of plesiosaur, but still similar to the one Mary discovered.

And new species of plesiosaurs—a most diverse group of aquatic carnivores—are being discovered to this day. One of the oldest and most complete skeletons of a prehistoric aquatic reptile has been uncovered in North America, representing an entirely new group of plesiosaurs. This 8.5-foot specimen, known as *Nichollsia borealis,* is one of the most complete and best-preserved North American plesiosaurs from the Cretaceous Period.

Much has been learned, too, about Mary's pterosaur, the repellent mutant that caused such a fright in the 1820s. Originally thought to be so bulky it could move only by gliding, today scientists believe the pterosaur was actually a nimble and athletic flyer, perhaps even able to outperform a modern bird. It is now known that these ancient reptiles, which flourished 251 to 65 million years ago, ranged in size from a sparrow to an airplane. In recent decades, pterosaur fossils have been discovered on every continent except Antarctica. The fossils of one massive pterosaur, known as *Quetzalcoatlus,* were first discovered in Texas in 1971. With a wingspan of up to 36 feet, it is among the largest flying animals ever found. At the other end of the spectrum, another new fossil species of pterosaur with a wingspan of less than 1 foot was discovered in China.

And what about Mary's first discovery, the ichthyosaur? At one time, scientists referred to it as the most ferocious reptile of its time. But today another newly discovered creature—the pliosaur—has replaced the ichthyosaur as the top marine predator. Often hailed as the *Tyrannosaurus rex* of the Jurassic sea world, the pliosaur is another species of extinct reptile that lived in the world's oceans during the age of dinosaurs. Pliosaurs had teardrop-shape bodies and two sets of powerful paddles that they used to "fly" through the water. Their short necks supported a massive skull sporting an impressive set of teeth, which they used to prey on squidlike animals, fish, and even other marine reptiles. Indeed, the Natural History Museum in Oslo, Norway, announced the discovery of one of the largest dinosaur-era marine reptiles ever found: an enormous pliosaur estimated to be almost 50 feet long.

But today there are many other land creatures that would make both the *T-rex* and the pliosaur look like runts. When *Giganotosaurus* was discovered in Argentina in the 1990s, with its 6-foot skull and 41-foot body, it was considered to be the largest meat eater known to humans. But then a new giant meat eater came along to replace it: *Mapusaurus,* discovered in Patagonia, Argentina, in 2006, was longer than 41 feet and weighed 3 tons. Then yet another creature, the plant-eating *Mamenchiasaurus,* came in at nearly 70 feet long, with half of that length being just its neck. Still another plant eater, the *Argentinosaurus,* stood three stories high and weighed as much as 50 elephants. But its head was no larger than that of a horse, and its brain was smaller than a human's. One of the most bizarre dinosaurs of all time has to be *Amargasaurus,* with spikes more than 2 feet tall protruding upward from its spine.

At present, over 700 different species of dinosaur have been identified and named, as well as countless other species of reptiles and other animals. Paleontologists believe that many more new and different species are still waiting to be found. In the central regions of Fiji, a new iguana species, the *Brachylophus bulabula,* has been discovered. It joins only two other living Pacific iguana species, one of which is critically endangered. Afraid of snakes? In early 2009, researchers announced the finding of 42-foot snake fossils in northeastern Colombia; the creature would have weighed more than a ton and could easily have eaten something the size of a cow. Scientists named it *Titanoboa cerrejonensis,* which means "titantic boa from Cerrejon," the region where it was found.

Perhaps most remarkable in this era of stunning archaeological advancements are the accompanying breakthroughs in genetics, biochemistry, and molecular biology that, through DNA sequencing, may soon give scientists the means by which to bring back some of the monstrous prehistoric creatures Mary and others introduced to the world.

Anyone who has seen any of the *Ice Age* movies is familiar with one of the main characters, the moody woolly mammoth named Manny voiced by actor Ray Romano. He wins sympathy from the audience as a loner-ish creature with a heart of gold. But in the eighteenth century, the idea that a woolly mammoth once existed raised the hackles of the religious establishment. When Georges Cuvier first proposed that mammoth bones found were those of an extinct form of elephant, many churchgoers went ballistic. But it soon turned out that Cuvier was right. Today, in one of the greatest scientific advancements of the early twenty-first century, the huge mammal that has been extinct for about 10,000 years is sparking controversy yet again. Researchers at Penn State University have sequenced about 80 percent of the woolly mammoth's gene map, using DNA taken from hair samples collected from a number of specimens. Will it be possible to bring the woolly mammoth back to life by inserting mammoth DNA sequences into the genome of the modern-day elephant? It's enough to make Cuvier turn over in his grave.

Mary's good friend William Buckland, too, would be shocked to learn that today, so many decades after his death, people are still spending their lives trying to find proof of the worldwide deluge of Noah's time, just as he had. The search for evidence goes on all around the world, from Mount Ararat in Turkey to the Elburz mountain range in Iran. Indeed, the last few decades have been a time of exciting discoveries in biblical archaeology, even if Noah's ark hasn't been found. Eric Cline, an archaeologist at George Washington University in Washington, D.C., with years of excavation experience in the Middle East, said in a conversation with the author that nothing so far has been found that would prove the existence of a worldwide deluge.

But Cline said that advancements in archaeology, such as magnetometers and precise excavation methods, have meant that at least some details in the Bible have been confirmed. For example, during the past century or so, archaeologists have found the first mention of Israel outside of the Bible, in an Egyptian inscription carved by the pharaoh Merneptah in the year 1207 BC. They have found mentions of Israelite kings, including Omri, Ahab, and Jehu, in neo-Assyrian inscriptions from the early first millennium BC. And most recently, they have found a mention of the House of David in an inscription from northern Israel dating to the ninth century BC. Some say that these are conclusive pieces of evidence that these people and places once existed and that at least parts of the Bible are historically accurate.

Even so, Cline warns that we are living in a time of widespread biblical fraud, dubious science, and crackpot theorizing. Some of the highest-profile discoveries of the recent past have been tainted by accusations of forgery, such as the James Ossuary, which may or may not be the burial box of Jesus' brother. Again and again, scientific expeditions embark on highly publicized journeys to search for proof of Noah's ark, raising untold amounts of money from believers who eagerly listen to tales spun by sincere amateurs or rapacious con men.

Meanwhile, the true and verifiable discoveries have been wielded in vigorous and ongoing debates over whether the Bible's account of events is meant to be taken literally—the same debate that was going on in Mary Anning's time.

To this day, real discoveries still are being made on the beaches of Lyme Regis. In 2000, David Sole, a former lawyer turned professional fossil collector, struck it big when he discovered a piece of a skeleton. Bit by bit, a spectacular, three-dimensionally preserved specimen emerged. It was a 185-million-year-old *Scelidosaurus,* the earliest of the armored dinosaurs, weighing half a ton.

There is no question that the fossil-hunting craze that started when Mary was alive is still going strong today. Every weekend, hordes of fossil hunters flock to the cliffs of southern England, a 95-mile stretch of shoreline now called the Jurassic Coast that was declared a UNESCO World Heritage Site in 2001. When a huge landslip occurred late in the evening on May 6, 2008—the worst in 100 years, destroying 1,300 feet of coast—crowds of fossil hunters gathered at the scene, just as Mary and William Buckland and so many others had nearly two centuries ago. And, like Mary, they were all looking for something special hidden amid fallen boulders, uprooted trees, and mounds of earth.

Sometimes fossil hunters risk their personal safety in their constant hope to make the discovery of a lifetime. Like Mary, they never give up.

Timeline

1799	Mary Anning is born in Lyme Regis, England.
1803	U.S. president Thomas Jefferson asks Meriweather Lewis and William Clark to explore the uncharted West.
1804	French anatomist Georges Cuvier suggests that some fossils are thousands of centuries old, a radical notion at the time.
1810	Mary Anning's brother, Joseph, discovers the skull of the world's first fossil ichthyosaur. Mary Anning will collect the whole skeleton the next year.
1819	The first steamship, the *Savannah*, crosses the Atlantic Ocean.
1823	Mary Anning finds the first complete *Plesiosaurus giganteus*.
1824	British geologist William Buckland publishes a paper on the *Megalosaurus* and thus describes for the first time a dinosaur fossil.
1824	British doctor and geologist Gideon Mantell publishes a paper on the *Iguanodon*, the second description of a dinosaur and the first description of an herbivorous fossil reptile.
1828	Mary Anning discovers the first British pterosaur, the first pterodactyl of the *Dimorphodon* genus.
1828	The first passenger railroad in the United States begins.
1829	Mary Anning discovers a new type of fossil fish, a *Squaloraja*.
1830	Mary Anning discovers a *Plesiosaurus macrocephalus*.
1830	British lawyer-turned-geologist Charles Lyell publishes *Principles of Geology*, a book that Charles Darwin will read while aboard the *Beagle*.
1830	To raise funds and to generate interest in Mary Anning's fossils, her friend and fellow geologist Henry De la Beche publishes a scene of marine life called "A More Ancient Dorset."

1831–1836	British scientist Charles Darwin sails on the *Beagle*, visiting, among other locations, the Galápagos Islands.
1832	British doctor and geologist Gideon Mantell finds the first fossil *Hylaeosaurus*, an ankylosaur.
1837	French naturalist Louis Agassiz shocks his audience by announcing his theory of an Ice Age.
1842	British anatomist Sir Richard Owen proposes the term *Dinosauria* ("terrible lizards").
1847	Mary Anning dies of breast cancer.
1859	British scientist Charles Darwin publishes his famous *On the Origin of Species by Means of Natural Selection*.

Notes

Much information on Mary Anning can be found in the writings of Professor Hugh Torrens, a leading expert on Anning. His research can be found in "Mary Anning of Lyme; The Greatest Fossilist the World ever knew," *British Journal of the History of Science*, vol. 28 (1995). Extensive information also can be found in the research of science historian William Lang (1878–1966). Details of her life can be found in his "Mary Anning of Lyme, Collector and Vendor of Fossils," *Natural History Magazine*, vol. 5, no. 34 (1936). Information about William Buckland's life can be found in a book by his daughter, Anna Gordon, *The Life and Correspondence of William Buckland* (London: John Murray, 1894). Buckland's inaugural lecture at Oxford, entitled "Vindiciae Geologicae," or "The Connexion between Geology and Religion Explained," was delivered on May 15, 1819 and published in 1820 in Oxford. Much information about the life of Richard Owen can be found in his grandson's biography: *Richard Owen, The Life of Richard Owen, Vol. 1* (London: John Murray, 1894).

1 Snakestones, Thunderbolts, and Verteberries

1. Hugh Torrens, "Presidential Address: Mary Anning (1799–1847) of Lyme; "The Greatest Fossilist the World Ever Knew," *British Journal of the History of Science* 28 (1995): 258.
2. John Fowles, *A Short History of Lyme Regis* (Wimborne, England: The Dovecote Press, 1991), p. 28.
3. Ibid.
4. Jane Austen, *Persuasion* (London: Penguin, 1998), p. 89.
5. Jane Austen, Letter 39 to Cassandra Austen, in *Jane Austen's Letters*, ed. Deidre Le Faye (Oxford: Oxford University Press, 1995).
6. Deborah Cadbury, *The Dinosaur Hunters* (London: Fourth Estate, 2000), p. 3.
7. Colin Dawes, *Fossil Hunting Around Lyme Regis* (Lyme Regis: Colin Dawes Studios, 2003), contains much information about Lyme Regis's fossils.
8. Information about Colyton comes from the Colyton Parish History Society at www.colytonhistory.co.uk.
9. Deborah Cadbury, *The Dinosaur Hunters* (London: Fourth Estate, 2000), p. 4.

10. *Bath Chronicle*, Dec. 27, 1798, p. 3.
11. Patricia Pierce, *Jurassic Mary* (Stroud, England: Sutton Publishing, 2006), p. 8.
12. George Roberts, *History of Lyme Regis & Charmouth* (London: Samuel Bagster, 1834), p. 286.
13. William Lang, a science historian who lived from 1878 to 1966, published many articles in the *Proceedings of the Dorset Natural History and Archeological Society*. One was "Mary Anning's Escape from Lightning" 80 (1959).
14. Roberts, *History of Lyme Regis & Charmouth*, p. 288.
15. J. M. Edmonds, "The Fossil Collection of the Misses Philpot of Lyme Regis," *Proceedings of the Dorset Natural History and Archeological Society* 98 (1976): 43.
16. From my own walks along the Dorset coastline.
17. W. D. Lang, "Mary Anning and the Pioneer Geologists of Lyme," *Proceedings of the Dorset Natural History and Archeological Society* 60 (1939).

2 A Fantastic Beast

1. Hugh Torrens, "Presidential Address: Mary Anning (1799–1847) of Lyme; The Greatest Fossilist the World Ever Knew," *British Journal of the History of Science* 28 (1995): 259.
2. George Roberts, *A Social History of the People of the Southern Counties of England in Past Centuries* (London: Longman, Brown, Green, Longman & Roberts, 1856), p. 562.
3. George Roberts, *History of Lyme Regis & Charmouth* (London: Samuel Bagster, 1834), p. 286.
4. Thomas W. Goodhue, *Curious Bones: Mary Anning and the Birth of Paleontology* (Greensboro, NC: Morgan Reynolds Publishing, 2002), p. 19.
5. Ibid., p. 17.
6. Roberts, *History of Lyme Regis & Charmouth*, p. 288.
7. Ibid.
8. E. P. Alexander, "William Bullock: Little-Remembered Museologist and Showman," *Curator* 28 (1985): 117–147.
9. Paul J. McCartney, *Henry De la Beche: Observations of an Observer* (Cardiff: Friends of The National Museum of Wales, 1977), pp. 2–7.

3 An Unimaginable World

1. *Dorchester and Sherborne Journal*, 48, Dorset County Museum.
2. Everard Home, "Some Account of the Fossil Remains of an Animal More Nearly Allied to Fishes than Any Other Classes of Animals," *Philosophical Transactions of the Royal Society* 571–577.

3. Charles Konig, *Synopsis of the Contents of the British Museum*, 11th ed. (London: 1817), p. 54.

4. George Bugg, *Scriptural Geology* (England: Hatchard and Son, 1826).

5. Simon Winchester, *The Map That Changed the World* (London: Viking, 2001), pp. 1–20.

6. Keith Thomson, *Before Darwin* (New Haven, CT: Yale University Press, 2005), pp. 187–188.

7. James Hutton, *Theory of the Earth with Proofs and Illustrations* (Edinburgh: Messrs Cadwell, Junior, Davies, and W. Creech, 1795), vol. 1, p. 3044.

8. Deborah Cadbury, *The Dinosaur Hunters* (London: Fourth Estate, 2000), pp. 22–24.

9. Much about Charles Peale and Thomas Jefferson can be found on the Academy of Natural Sciences in Philadelphia's Web site at www.ansp.org.

10. Thomas Jefferson, "Notes on the States of Virginia," Boston, Part of the Coolidge Collection of Thomas Jefferson Manuscripts at the Massachusetts Historical Society, 1781.

11. Christopher McGowan, *The Dragon Seekers* (London: Little, Brown, 2002), p. 5.

12. Cadbury, *Dinosaur Hunters*, p. 13.

13. Cadbury, *Dinosaur Hunters*, p. 21.

14. Anna Gordon, *The Life and Correspondence of William Buckland* (London: John Murray, 1894), pp. 33–39.

15. Cadbury, *Dinosaur Hunters*, pp. 19–20.

4   A Great Kindness

1. Mary Anning's #4 notebook, *Lang Papers*, Dorset County Museum, Dorchester, England.

2. Anonymous pamphlet was called "Catalogue of the Pictures Painted by the Late Benjamin West, Including a Description of the Great Pictures Christ Rejected and Pale Horse." It is not known when Mary copied this pamphlet or when it was published exactly.

3. Journals of Anna Maria Pinney, 1831–1833, quoted in W. D. Lang, "Mary Anning and Anna Maria Pinney," *Proceedings of the Dorset Natural History and Archeological Society* 76 (1956): 146–152.

4. Christopher McGowan, *The Dragon Seekers* (London: Little, Brown, 2002), p. 18.

5. Ibid., pp. 22–23.

6. Journals of Anna Maria Pinney, quoted in Lang, "Mary Anning and Anna Maria Pinney."

7. John Murray, "The Late Miss Mary Anning" *Mining Journal* 17 (1847).

8. Thomas Hawkins, *Memoirs of Ichthyosauri and Plesiosauri* (London: Relfe & Fletcher, 1834), pp. 9 and 26.

9. Paul J. McCartney, *Henry De la Beche: Observations of an Observer* (Cardiff: Friends of The National Museum of Wales, 1977), pp. 6–7.

10. Ibid., pp. 12–20.

11. J. M. Edmonds, "The Fossil Collection of the Misses Philpot of Lyme Regis," *Proceedings of the Dorset Natural History and Archeological Society* 98 (1976): 43.

12. Ibid., p. 47.

13. Ibid.

14. William Buckland's inaugural lecture, "*Vindiciae Geologicae*" (Oxford, 1820).

15. William Paley, *Natural Theology* (Oxford: Oxford University Press, 2008 [new edition]).

16. Patricia Pierce, *Jurassic Mary* (Stroud, England: Sutton Publishing, 2006), p. 146.

17. Hugh Torrens, *Newsletter of the Geological Curators Group* 2 (1979): 409.

18. Hugh Torrens, "Mary Anning's Life and Times: New Perspectives," Mary Anning Symposium, 1999; Lyme Regis, June 2–4, 99.

19. Hugh Torrens, "Presidential Address: Mary Anning (1799–1847) of Lyme; The Greatest Fossilist the World Ever Knew," *British Journal of the History of Science* 28 (1995): 261.

5 A Long-Necked Beauty

1. Charles Dickens, "All the Year Round," his weekly literary journal, Feb. 11, 1865 ed., p. 62.

2. Thomas W. Goodhue, *Curious Bones: Mary Anning and the Birth of Paleontology* (Greensboro, NC: Morgan Reynolds Publishing, 2002), p. 34.

3. Patricia Pierce, *Jurassic Mary* (Stroud, England: Sutton Publishing, 2006), p. 200.

4. W. D. I. Rolffe, A. C. Milner, and F. G. Hay, "The Price of Fossils," *Special Papers in Paleontology* 40 (1988): 149, quoted in Torrens, "Presidential Address," p. 262.

5. *Bristol Mirror*, Jan. 11, 1823, p. 4.

6. Goodhue, *Curious Bones*, p. 39.

7. W. D. Lang, "Mary Anning and the Pioneer Geologists of Lyme," pp. 152–153.

8. Pierce, *Jurassic Mary*, pp. 122–123.

9. Geological Society meeting minutes, February 1824, pp. 404–412.

10. Lang, "Mary Anning and the Pioneer Geologists," p. 153.

11. Goodhue, *Curious Bones*, p. 40.

12. Ibid.

13. *Transactions of the Geological Society of London* (1824): 390–396.

14. Deborah Cadbury, *The Dinosaur Hunters* (London: Fourth Estate, 2000), pp. 108–109.

15. Pierce, *Jurassic Mary*, p. 103.

16. Thomas Allan, "Travels in England," unpublished journal, quoted in Lang, "Mary Anning and the Pioneer Geologists," p. 154.

17. E. Welch, "Lady Silvester's Tour through Devonshire in 1824," *Devon and Cornwall Notes and Queries* 30 (1967): 313; 32 (1967): 265–266.

18. The references to Frances Bell are from Grant Johnson, *Memoir of Miss Frances Augusta Bell* (London: Hatchard & Son, 1827).

19. George Roberts, *History of Lyme Regis & Charmouth* (London: Samuel Bagster, 1834), p. 174.

20. Cited on the site www.discoveringfossils.co.uk.

6   The Hidden Mysteries of Coprolites

1. George Roberts, *History of Lyme Regis & Charmouth* (London: Samuel Bagster, 1834), p. 180.

2. Ibid., p. 178.

3. Ibid., pp. 178–179.

4. Ibid., pp. 258–260.

5. W. D. Lang, "Mary Anning and Anna Maria Pinney," pp. 146–152.

6. Patricia Pierce, *Jurassic Mary* (Stroud, England: Sutton Publishing, 2006), pp. 96–97.

7. E. B. and D. S. Berkeley, *George William Featherstonhaugh, the First U.S. Government Geologist* (Tuscaloosa, AL: University of Alabama, 1988), quoted in Hugh Torrens, "Presidential Address," p. 265.

8. Lang, "Mary Anning and the Pioneer Geologists," p. 55.

9. W. D. Lang, "Three Letters of Mary Anning," p. 171.

10. Information about De la Beche's life comes from Paul J. McCartney, *Henry De La Beche: Observations of an Observer* (Cardiff: Friends of National Museum of Wales, 1977), pp. 25–27.

11. George Roberts, *History of Lyme Regis & Charmouth* (London: Samuel Bagster, 1834), p. 325.

12. Nicolaas Rupke, *The Great Chain of History: William Buckland and the English School of Geology (1814–1849)* (Oxford: Oxford University Press, 1983), p. 142.

13. W. D. Lang, "Three Letters of Mary Anning," p. 169.

14. William Buckland, "Fossil Sepia," *London and Edinburgh Philosophical Magazine* 5 (1829): 388.

15. J. M. Edmonds, "The Fossil Collection of the Misses Philpot of Lyme Regis," *Proceedings of the Dorset Natural History and Archeological Society* 98 (1976): 45.
16. *Transactions of the Geological Society of London,* 2nd series, 3 (1829).

7 Finally, the Big City of London

1. All Frances Bell references are from Grant Johnson, *A Memoir of Miss Frances Augusta Bell* (London: Hatchard & Son, 1827), pp. 117–122, 129–136.
2. *Transactions of the Geological Society of London,* 2nd series, no. 3 (1829): 217–222, quoted in George Roberts, *History of Lyme Regis & Charmouth* (London: Samuel Bagster, 1834), pp. 325–326.
3. Thomas W. Goodhue, *Curious Bones: Mary Anning and the Birth of Paleontology* (Greensboro, NC: Morgan Reynolds Publishing, 2002), pp. 57–58.
4. Gideon Mantell, "Age of Reptiles," *Edinburgh New Philosophical Journal* 2 (1831): 181–185.
5. Elizabeth Oke Buckland Gordon, *The Life and Correspondence of William Buckland, DD, FRS, Sometime Dean of Westminster, Twice President of the Geological Society, and First President of the British Association* (London: John Murray, 1894), p. 27.
6. Charles Lyell, *Principles of Geology* (London: Penguin Group, 1997), p. xxiv.
7. Charles Lyell's *Principles of Geology* was published by John Murray in three volumes between 1830 and 1833.
8. *Salisbury & Winchester,* March 2, 1829, and May 4, 1829.
9. George Cumberland, *Royal Institution of Great Britain's Quarterly Journal of Literature, Science, and the Arts* 27 (1829): 348.
10. Paul J. McCartney, *Henry De la Beche: Observations of an Observer* (Cardiff: Friends of the National Museum of Wales, 1977), p. 27.
11. W. D. Lang, *Dorset Natural History and Archaeological Society Proceedings* 66 (1945): 170.
12. All references to the London Diary, Richard Owen Correspondence, General Library, Natural History Museum, vol. 1. The diary was purchased by Lord Enniskillen from Mary's nephew, Albert Anning.
13. Some information about London's history was found at www. victorianweb.org.

8 An Amazing New Fish

1. W. D. Lang, "Three Letters of Mary Anning."
2. Ibid.
3. *Salisbury & Winchester,* December 28, 1829.

4. Michael Taylor and Hugh Torrens, *Dorset Natural History and Archeological Society Proceedings* 108 (1986): 136, 142.

5. Hugh Torrens and M. Taylor, "Saleswoman to a New Science: Mary Anning the Fossil Fish Squaloraja from the Lias of Lyme Regis," *Proceedings of the Dorset Natural History and Archeological Society* 108 (1986): 140.

6. George IV's deathbed declaration is available from various sources, including www.family-ancestry.co.uk.

7. *New Brighton Guide*, 1796.

8. Thomas W. Goodhue, *Curious Bones: Mary Anning and the Birth of Paleontology* (Greensboro, NC: Morgan Reynolds Publishing, 2002), p. 50.

9. There is much information about "Duria Antiquior" and on De la Beche's other artwork in Paul J. McCartney, *Henry De la Beche: Observations of an Observer* (Cardiff: Friends of The National Museum of Wales, 1977).

10. Goodhue, *Curious Bones*, p. 62.

11. W. D. Lang, "Mary Anning and the Pioneer Geologists," p. 155.

12. Goodhue, *Curious Bones*.

13. Patricia Pierce, *"Jurassic Mary"* (Stroud, England: Sutton Publishing, 2006), pp. 96–97.

14. Information on the Church of England's history is available from James Bettley, an architectural historian, at www.vam.ac.uk.

9 Spilling Secrets

1. Much about Anna Maria Pinney has been included in W. D. Lang, "Mary Anning and Anna Maria Pinney." Anna Maria Pinney's diaries are at the Bristol University Library.

2. Patricia Pierce, *Jurassic Mary* (Stroud, England: Sutton Publishing, 2006), p. 67.

3. Thomas W. Goodhue, *Curious Bones: Mary Anning and the Birth of Paleontology* (Greensboro, NC: Morgan Reynolds Publishing, 2002), p. 72.

4. Much about Hawkins is from Thomas Hawkins, *Memoirs of Ichthyosauri and Plesiosauri: Extinct Monsters of the Ancient Earth and Twenty-One Plates Copied from Specimens in the Author's Collection* (London: Relfe & Fletcher, 1834), pp. 12–13, 25–28.

5. W. D. Lang, "Three Letters of Mary Anning," pp. 169–173.

6. J. Murray, "The Late Miss Mary Anning," *Mining Journal* 17 (1847): 591.

7. Pinney diaries.

8. Goodhue, *Curious Bones*, pp. 73–74.

9. Marilyn Bailey Ogilvie and Joy Dorothy Harvey, *The Biographical Dictionary of Women in Science* (Routledge, 1999), p. 202.

10. Pierce, *Jurassic Mary*, pp. 144–145.
11. Details about Queen Victoria are from Helen Rappaport, *Queen Victoria: A Biographical Companion* (ABC-CLIO, 2003).

10   Esteemed Visitors

1. W. D. Lang, *Dorset Proceedings* 66 (1945): 169–173; *Dorset Proceedings* 71 (1950): 184–188.
2. Buckland letter dated December 9, 1833, Oxford University Museum of Natural History.
3. Deborah Cadbury, *The Dinosaur Hunters* (London: Fourth Estate, 2000), p. 167.
4. Information about Charles Lyell and disagreements over Earth's origin is from Keith Thomson, *Before Darwin* (New Haven, CT: Yale University Press, 2005), pp. 139 and 186.
5. Martin J. S. Rudwick, *Scenes from Deep Time* (Chicago: University of Chicago Press, 1995), p. 48.
6. W. D. Lang, "Mary Anning and a Very Small Boy."
7. Christopher McGowan, *The Dragon Seekers* (London: Little, Brown, 2002), p. 179.
8. Details about Louis Agassiz are from Agassiz, *Louis Agassiz: His Life and Correspondence* (Whitefish, MT: Kessinger Publishing, 2004), pp. 1–18.
9. Thomas W. Goodhue, *Curious Bones: Mary Anning and the Birth of Paleontology* (Greensboro, NC: Morgan Reynolds Publishing, 2002), p. 81.
10. Gideon Mantell, *The Journal of Gideon Mantell, Surgeon and Geologist* (London: Oxford University Press, 1940), p. 108.
11. Goodhue, *Curious Bones*, p. 83.
12. Patricia Pierce, *Jurassic Mary* (Stroud, England: Sutton Publishing, 2006), p. 182.
13. Mary Anning's letter, April 7, 1839, *Magazine of Natural History* (December 1839).
14. W. D. Lang, "Mary Anning and the Pioneer Geologists," p. 159.
15. Ibid.

11   The Earth Moves

1. W. D. Lang, "Mary Anning and the fire at Lyme in 1844," *Proceedings of the Dorset Natural History and Archeological Society* 74 (1953), p. 175.
2. Christopher McGowan, *The Dragon Seekers* (London: Little, Brown, 2002), p. 182.
3. Thomas W. Goodhue, *Curious Bones: Mary Anning and the Birth of Paleontology* (Greensboro, NC: Morgan Reynolds Publishing, 2002), p. 90.

4. Patricia Pierce, *Jurassic Mary* (Stroud, England: Sutton Publishing, 2006), p. 161.

5. Robert Southey, ed., *The Complete Works of Henry Kirke White* (Boston: Whitaker, 1831). Located in No. 4 notebook in the Lang Papers.

6. No. 4 notebook in the Lang Papers.

7. Information about Richard Owen can be found at the University of California Museum of Paleontology's site at www.ucmp. berkeley.edu.

8. www.victorianweb.org.

9. Much information about De la Beche can be found in Paul J. McCartney, *Henry De la Beche: Observations of an Observer* (Cardiff: Friends of the National Museum of Wales, 1977).

10. Goodhue, *Curious Bones*, p. 93.

11. W. D. Lang, "Mary Anning and the fire at Lyme in 1844," *Proceedings of the Dorset Natural History and Archeological Society* 74 (1953), p. 175.

12. C. G. Carus, *The King of Saxony's Journey through England and Scotland in the Year 1844* (London: Chapman & Hall, 1846), p. 197.

12 The Making of a Legend

1. Patricia Pierce, *Jurassic Mary* (Stroud, England: Sutton Publishing, 2006), p. 183.

2. Thomas W. Goodhue, *Curious Bones: Mary Anning and the Birth of Paleontology* (Greensboro, NC: Morgan Reynolds Publishing, 2002), pp. 93–94.

3. Much about Mary's last few years and her death were written about by W. D. Lang in "Mary Anning of Lyme, Collector and Vendor of Fossils," *Natural History Magazine* 5, no. 34 (1936). There is also much written in Hugh Torrens, "Mary Anning of Lyme; the Greatest Fossilist the World Ever Knew," *British Journal of the History of Science* 28 (1995).

4. References to the rhyme can be found in many places, including the *Oxford Dictionary of Quotations* and the site www.discoveringfossils. co.uk.

5. Charles Dickens, *All the Year Round*, his weekly literary journal, vol. 13 (1965), pp. 60–63.

6. George Roberts, *History of Lyme Regis & Charmouth* (London: Samuel Bagster, 1834), p. 290.

7. W. D. Lang, "Mary Anning and the Pioneer Geologists."

8. Peter Watts, "Seven Wonders of London: Natural History Museum." http://www.timeout.com/london/big-smoke/features/3893/Seven_wonders_of_London-Natural_History_Museum.html.

9. The Natural History Museum's Web site at www.nhm.ac.uk.

10. Comments on the Natural History Museum made by H.V. Morton in "In Search of London," 1951, London, Methuen Publishing, as cited by Peter Watts, "Seven Wonders of London: Natural History Museum." http://www.timeout.com/london/big-smoke/features/3893/Seven_wonders_of_London-Natural_History_Museum.html.

Index